FAILURE CASE STUDIES IN CIVIL ENGINEERING

STRUCTURES, FOUNDATIONS, AND THE GEOENVIRONMENT

SECOND EDITION

EDITED BY

Paul A. Bosela, Ph.D., P.E.
Pamalee A. Brady, Ph.D., P.E.
Norbert J. Delatte, Ph.D., P.E.
M. Kevin Parfitt, P.E.

SPONSORED BY

Technical Council on Forensic Engineering
of the American Society of Civil Engineers

American Society of Civil Engineers

Cataloging-in-Publication Data on file with the Library of Congress.

Published by American Society of Civil Engineers
1801 Alexander Bell Drive
Reston, Virginia, 20191-4400
www.asce.org/pubs

Contents

Preface v
Acknowledgments vii

1 . **Foundation Failures**...1
Tower of Pisa (1173 & Ongoing) 2
Transcona Grain Elevator (1913) 5
Fargo Grain Elevator (1956)...7
La Playa Guatemala Earthquake (1976)...8
Schoharie Creek Bridge (1987) 9

2. **Embankment, Dam, and Slope Failures** 14
St. Francis Dam (1928)..15
Malpasset Dam (1959)...17
Vajont Dam (1963) 19
Lower San Fernando Dam (1971) 22
Teton Dam (1976)...24
Rissa Norway Landslide (1978) 29
Nerlerk Berm Failure (1983) 30
Carsington Embankment (1984)..32

3. **Geoenvironmental Failures**..33
Love Canal (1978) 34
Valley of the Drums (1978)..36
Stringfellow Acid Pits (1980)...38
Seymour Recycling Facility (1980)...40
Kettleman Hills Waste Landfill (1988) 42
North Battleford, Saskatchewan Water Treatment Failure (Cryptosporidium
    Outbreak) (2001)..45

4. **Bridge Failures**...49
Ashtabula Bridge (1876)..50
Tay Bridge (1879)...52
The Quebec Bridge (1907 & 1916) 54
Falls View Bridge (1938) 56
Sando Arch (1939)...57
Tacoma Narrows Bridge (1940) 58
Peace River Bridge (1957)..61
The Second Narrows Bridge (1958) 62
King Street Bridge (1962)...63
Point Pleasant Bridge—Silver Bridge (1967)..64
Antelope Valley Freeway Interchange (1971 & 1994)..66
Mianus River Bridge (1983)...69
San Francisco-Oakland Bay Bridge (1989) 71
Cypress Viaduct (1989) 74

5. Building Failures..**76**
   AMC Warehouse (1955)...77
   Ronan Point Tower (1968) 79
   2000 Commonwealth Avenue (1971)..82
   The Skyline Plaza Apartment Building (Bailey's Crossroads) (1973)............85
   Hartford Civic Center Coliseum (1978) 88
   Imperial County Services Building (1979)...93
   Kemper Memorial Arena Roof (1979) 95
   Binghamton State Office Building (1981)..98
   Hyatt Regency Hotel Pedestrian Walkways (1981) 100
   Pino Suarez Building (1985) 104
   L'Ambiance Plaza (1987)...106
   Burnaby Supermarket Rooftop Parking Deck (1988)............................110
   Northridge Meadows Apartments (1994)...112
   California State University, Northridge, Oviatt Library (1994)......................114
   Alfred P. Murrah Federal Building (1995)...116
   Charles De Gaulle Airport Terminal 2E (2004) 118
   Four Times Square Scaffold Collapse (Conde Nast Tower) (1998)..................120

Index 123

Preface

This is a special publication of the Education Committee of the American Society of Civil Engineers' (ASCE) Technical Council on Forensic Engineering (TCFE). It was first published in 1995 as *Failures in Civil Engineering: Structural, Foundation and Geoenvironmental Case Studies* edited by Robin Shepherd and J. David Frost. Forensic Engineering is the application of engineering principles to the investigation of failures or other performance problems. The investigations may involve testimony on the findings before a court of law or other judicial forum, when required. Failures include not only catastrophic events, such as bridge and building collapses, but also failures of facilities or components to perform as intended by the owner, design professional, or constructor. ASCE authorized TCFE in July 1985 following a number of dramatic collapses of engineered structures, such as the Hartford Coliseum, Kemper Arena, and the Hyatt Regency Walkways. The purpose of TCFE is to develop practices and procedures to reduce the number of such failures, to disseminate information on failures and their causes, to provide guidelines for conducting failure investigations, and to provide guidelines for ethical conduct in forensic engineering. It is the purpose of TCFE's Education Committee to promote the study of failure case histories in educational activities. Thus, the committee works to promote and advance the educational objectives of colleges and universities and act as a source of referral for educational material with forensic engineering emphasis.

Design and construction of civil engineering projects presents unique challenges. The design and construction "team" consists of owners, design professionals, and construction professionals in a temporary association for a specific project. The unpredictability of the exposure to natural and man-made hazards that the facility will experience during its life necessitates that the design be based on sound engineering judgment rather than certainty. In some industries, a series of prototypes can be designed and built to work out the problems before the final product is manufactured. Lessons learned from problems with the earlier prototypes are incorporated in the final product. By the nature of most civil engineering projects, that is not an option. Hence, it is critical that we learn from both the successes and failures of each individual project.

Failures can occur during construction, any time during the service life of the facility, and even during demolition. They occur for many reasons, such as inadequate consideration of the inherent instabilities of a structure, fragmentation of responsibilities that may occur during the construction process, not anticipating or designing for all modes of failure, and unusual or unanticipated loadings. Failures can often be traced to the beginning of a project's life; such errors include:

- Design errors and omissions
- Construction sequencing (means and methods)
- Construction phase process failures (e.g. shop drawings and submittals)
- Construction defects
- Materials defects
- System and component defects

Once a building or structure is in service, failure not directly attributable to design or construction is most often caused by one or more of the following:

- Deterioration
- Damage
- Catastrophic events
- Overload

Failures often result from a combination of factors, and it is instructive to study the underlying cause(s), triggering events, contributing factors, and mitigating factors. Sometimes, one failure results in a change in the standard of practice. Other times, it takes a series of repeated failures to influence change. Unfortunately, despite repeated failures, the lessons are not always learned. This second edition has been expanded with the inclusion of additional case studies such as the Alfred P. Murrah Building and Charles de Gaulle Airport Terminal as well as expanded descriptions of the case studies, and inclusion of additional references, photographs, and illustrations.

The objective of this publication is to provide resource material on typical failure case studies that can be used by engineering professors in their classrooms. It includes an outline of each case study, summary of the lessons learned, and a list of references for further study. It is anticipated that the discussion of failure case studies in the classroom will provide students with improved awareness that sound engineering requires an understanding of both the successes as well as the failures.

Acknowledgments

The Education Committee wishes to acknowledge the contributions of those involved in this publication. The original publication included short study descriptions contributed by the following individuals:

- Paul A. Bosela, Cleveland State University, Cleveland, OH
- Kenneth L. Carper, Washington State University, Pullman, WA
- Timothy J. Dickson, Construction Technology Laboratories, Inc., Skokie, IL
- J. David Frost, Georgia Institute of Technology, Atlanta, GA
- Narbey Khachaturian, University of Illinois, Champaign-Urbana, IL
- Oswald Rendon-Herrero, Mississippi State University, Starkville, MS
- Robin Shepherd, Forensic Expert Advisers, Inc., Santa Ana, CA

The following individuals contributed new case studies and other materials for the second edition:

- Paul A. Bosela, Cleveland State University, Cleveland, OH
- Pamalee A. Brady, California Polytechnic State University, San Luis Obispo, CA
- Kenneth L. Carper, Washington State University, Pullman, WA
- Norbert J. Delatte, Cleveland State University, Cleveland, OH
- M. Kevin Parfitt, Penn State University, University Park, PA
- Kevin Rens, The University of Colorado Denver, Denver CO
- Kevin Sutterer, Rose-Hulman Institute of Technology, Terre Haute, IN

The committee also wishes to acknowledge the contribution of Pennsylvania State University student Gaby Issa-El-Khoury, who contributed to the research on the Charles de Gaulle Airport Terminal 2E case history and provided summary translations of the original French investigation report and other articles. The Technical Council of Forensic Engineering Executive Committee provided review and support of this second edition. The Education Committee thanks Executive Committee members David B. Peraza (Chairman), Leonard J. Morse-Fortier (Past-Chair), Anthony M. Dolhon (Vice-Chairman), Michael P. Lester (Secretary), Michael J. Drerup, and Glenn G. Thater.

Paul A. Bosela
Pamalee A. Brady
Norbert J. Delatte
M. Kevin Parfitt
Editors

Chapter 1

Foundation Failures

TOWER OF PISA
(1173 & Ongoing)

The Tower of Pisa (Figure 1-1) in Italy is about 60 m (200 ft) tall from foundation to belfry, 20 m (66 ft) in diameter and weighs approximately 145 MN (14,500 tons). It was constructed in three phases. Four floors were built over a 5 year period from 1173 to 1178. Following an almost 100 year hiatus, in the second phase of construction, three additional floors were constructed between 1272 and 1278. The third construction phase occurred more than 80 years later between 1360 and 1370 when the bell tower was added. The tower's foundation is inclined at almost 5.5 degrees to the south; the tower overhangs the ground about 6 m (20 ft) out of plumb. The value corresponding to the eccentricity on the loads on the foundation is 2.3 m (7.5 ft).

Evidence indicates that the phased construction was mandated by the performance of the structure as it was being built. Further, records indicate that during construction the tower appeared to move sufficiently so that the builders used obliquely cut stones in an effort to maintain the floor of each successive story approximately horizontal. It is interesting to note that the obliquely cut stones were used by the Pisa Commission, entrusted with gathering and collating relevant data for an international competition organized to identify a method to stabilize the tower in 1972, to reconstruct the pattern of movements throughout the first two phases of the tower's construction. Their calculations showed that at the end of the first construction phase the tower had begun to lean towards the northwest. During the second and third phases, the angle of inclination increased and the principal direction of tilt shifted first to the northeast and then to the south.

By 1993 the tower's maximum horizontal tilt had progressed to 5.2 meters (17 ft). In June of that year an attempt was made to reduce the tower's tilt. Specially fabricated lead counterweights were placed on top of the north side of a tensioned concrete ring built around the base of the tower. The tilt was reduced by approximately 12 seconds of arc over a 6-week period, as about 1.3 MN (130 tons) of lead were placed. Over the course of the remediation, until January 1994, a total of 6.9 MN (690 tons) was applied and by July 1994 the tower had righted its position toward the north a full 52 arc seconds. The Pisa Commission decided to replace the lead counterweights with an anchored cable system. In 1995 they began freezing the ground with liquid nitrogen in preparation for installing the cables. As soon as the freezing stopped the tower began to once again lean southward at a rate of four arc seconds per day. The operation was halted. They continued a search for a permanent solution.

In March of 1996 engineers successfully completed a test of a soil-extraction method to reduce the tower's lean. In this method an inclined drill was used to create cavities that gently closed due to the pressure of the overlying soil. The method was fully employed three years later. In 1999, after the installation of temporary cables, which could be tensioned to steady the tower if detrimental movements occurred, engineers drilled a dozen boreholes over a width of approximately 5.5 m (18 ft). They slowly removed underlying soil at a rate of approximately 0.02 cubic meters (five gallons) every two days. By the end of August the lean had decreased by 130 arc seconds (38 mm or 1.5 in.). The success of the method allowed for the removal of three lead counterweights. The full intervention involved the drilling of 41 tubes over the entire width of the tower causing righting movement of about 6 seconds of arc per day. Lead counterweights began to be removed in May 2000 and continued until the final counterweight was removed in January 2001 while soil extractions continued. The concrete ring on which the weights had sat was also removed and in March 2001 augers and casing were removed and the bore holes filled with a bentonitic grout. A small rotation was associated with the removal of the cables; slight soil extraction occurred in June 2001. The tower was reopened to the public in December 2001.

The first recorded modern direct measurement of the angle of inclination was in 1911. Measurements made since then have shown that when no external forces/conditions are acting, the rate of inclination increases approximately at a constant rate. For example, from mid-1975 to mid-1985 when external disturbances were at a minimum, the average rate of inclination remained approximately constant at about 5.8 seconds of arc per year. Conversely, drawdown of groundwater during the period late 1970 to late 1974 caused a dramatic increase in the inclination rate to 11.9 seconds of arc per year. Similar increases have accompanied other invasive construction activities over the past 80 years.

Lessons Learned

The failure of the Tower of Pisa is unique for a number of reasons. It is a failure that has occurred on a continuous basis for more than 800 years. Despite the extensive investigations and analyses conducted over the past 60 years, there is still no consensus on the cause of failure. What is significant however is that finally, after 8 centuries, the tower has been effectively stabilized.

References

Burland, J. (2002). "Preserving Pisa's Treasure," *Civil Engineering*, March, 72.

Jamiolkowski M. (1994). "Leaning Tower of Pisa - Description of the Behavior," *Settlement '94 Banquet Lecture*, College Station, TX, 55.

Leonards, G.A. (1979). Discussion of "Foundation Performance of the Tower of Pisa," *Journal of Geotechnical Engineering Division*, 105(GT1), 95-105.

Mitchell, J., K., Vivatrat, V., and Lambe, T.W. (1977). "Foundation Performance of the Tower of Pisa," *Journal of Geotechnical Engineering Division*, 103(GT3), 227-247.

NOVA. (1999). *Fall of the Leaning Tower*, originally broadcast October 5, NOVA. Available at http://www.pbs.org/wgbh/nova/pisa/.

Figure 1-1. Tower of Pisa.
Courtesy of Paul J. Parfitt (Wiss, Janney, Elstner Associates, Inc.).

TRANSCONA GRAIN ELEVATOR
(1913)

Construction of a million-bushel grain elevator began in 1911 at North Transcona (near Winnipeg), Manitoba, Canada. The elevator structure consisted of a reinforced concrete work-house and an adjoining bin-house. In plan, the work-house measured 21 by 29 m (70 by 90 ft). The structure was 55 m (180 ft) high and was founded on a raft foundation 3.6 m (12 ft) below grade. The bin-house consisted of five rows of thirteen bins each 4.3 m (14 ft) in diameter and 28 m (92 ft) high that rested on a concrete framework supported by a reinforced concrete raft. The bin-house raft measured 23 m (77 ft) by 59 m (195 ft) and was also founded at a depth of 3.7 m (12 ft) below grade.

Upon completion in September 1913 filling of the elevator commenced and the grain was distributed as evenly as practicable to the bins. On October 18, 1913, after 31,500 cu m (875,000 bushels) of wheat had been placed in the elevator settling was noted and increased uniformly to about 0.3 m (1 ft) per hour. Tilting of the elevator then began to occur; it ceased 24 hours later when the inclination reached 26 degrees 53 minutes from vertical.

The subsoil below the elevator's foundation consisted of a uniform deposit of clay that resulted from the sedimentation in waters of glacial Lake Aggassiz. The clay was a varved, slickensided, highly plastic material varying from about 9 to 15 m (30 to 50 ft) in depth. It overlaid glacial till over limestone bedrock. From the ground surface to a depth of about 9 m (30 ft), the clay had a stiff consistency that gradually decreased with a decrease in depth. A water table was encountered at a depth of about 9 m (30 ft).

In 1951, a comprehensive geotechnical investigation led to the conclusion that the elevator foundation failed due to a bearing failure in the underlying clay. Unfortunately, for the design engineers, the state-of-the-art in geotechnical engineering in 1911 had not reached the point that the ultimate bearing capacity could be computed. No borings were known to have been made for the design of the elevator.

Lessons Learned

The development of soil mechanics after the Transcona failure eventually provided a basis for computing the ultimate bearing capacity of soils. It was subsequently realized, therefore, that the Transcona failure served as a "full-scale" check of the validity of such computations. In hindsight, had the Transcona engineers had access to soil mechanics theory, the failure could have been averted.

References

Allaire, A. (1916). "The Failure and Righting of a Million-Bushel Grain Elevator," *ASCE Transactions*, LXXX, 800-803.

Peck, R.B. and Bryant, F.G. (1953). "The Bearing-Capacity Failure of the Transcona Elevator," *Geotechnique*, III, 201-208.

White, L.S. (1953). "Transcona Elevator Failure: Eye-Witness Account," *Geotechnique*, III, 209.

FARGO GRAIN ELEVATOR
(1956)

Beginning in the summer of 1954, a reinforced concrete grain elevator was constructed on level field near Fargo, North Dakota. During the autumn and the winter of 1954, a small amount of grain was placed in the elevator. Major filling of the elevator did not begin until the latter part of April 1955. On the morning of June 12, 1955, the elevator experienced a classic, full-scale bearing capacity failure, and was completely destroyed.

The grain elevator was a reinforced concrete structure consisting of twenty circular bins, twenty-six small interstitial bins at one end, and a combined bin and work-house at the other end of the structure. In plan, the structure is a long rectangular area measuring 16 by 67 m (52 by 218 ft) overall. The structure's foundation was a reinforced concrete raft of 0.7 m (2.3 ft) in thickness. The bottom of the raft foundation was located 1.8 m (6 ft) below grade. Sheet piles lined the raft periphery and were thought to be driven to a depth of 5.5 m (18 ft). The elevator was founded on the fine grained sediments of Old Lake Aggassiz.

A study including a subsoil investigation and laboratory testing was conducted to determine if the failure had resulted from inadequate bearing capacity of the subsoil. The results of the study indicated that overstressing of the subsoil was the reason for the failure. Investigation of full-scale foundation failures are rare; like the Transcona grain elevator failure, this case also affords the opportunity to assess the state-of-the-art in bearing capacity analysis.

Lessons Learned

The Fargo grain elevator collapse provides a full-scale example of a textbook bearing capacity failure. The failure provided unique and useful bearing capacity data for assessing the validity of state-of-the-art analytical procedures.

References

Deere, D.U., and Davisson, M.T. (1965). " Behavior of Grain Elevator Foundations Subjected to Cyclic Loading," *Fifth International Conference on Soil Mechanics and Foundation Engineering.*

Delatte, Norbert J. (2009). *Beyond Failure: Forensic Case Studies for Civil Engineers*, ASCE Press, Reston, VA, 251-255.

Morley, J. (1996). "'Acts of God': The Symbolic and Technical Significance of Foundation Failures," *Journal of Performance of Constructed Facilities*, 10(1), 23 – 31.

Norlund, R.L. and Deere, D.U. (1970). "Collapse of Fargo Grain Elevator," *Journal of the Soil Mechanics and Foundation Engineering Division*, 96(SM2), 585-607.

LA PLAYA GUATEMALA EARTHQUAKE
(1976)

On February 4, 1976, an earthquake of 7.5 magnitude on the Richter scale devastated the vacation community of La Playa, Lake Amatitlan, Guatemala. The earthquake induced liquefaction in the subsoil underlying La Playa. Its epicenter was located 170 km (106 miles) northeast of Lake Amatitlan. Within the zone of heavy damage, there was subsidence and flooding of beach areas and severe ground cracking that resulted in severe damage to houses as well as numerous sand boils. Of 32 houses at La Playa, 29 were destroyed or damaged, generally as the result of lateral ground spreading and subsidence. Most of the vacation houses were unoccupied at the time of the earthquake. A comprehensive study of the earthquake and its effects was conducted including field and laboratory sampling and testing.

The results of the study provided a very useful case history in which field data on soil characteristics in an earthquake-liquefied zone and a non-liquefied zone could be correlated with field performance. These results supplemented the limited number of available case studies of this type that could be used for the predictions of probable behavior at other sites and permitted corroboration of analytical procedures.

Lessons Learned

The results of the field studies from La Playa contribute to the available database relating to earthquake-induced liquefaction and thus improve the predictive capability in earthquake-prone regions.

References

Espinosa, A.F. (1976). *The Guatemala Earthquake of February 4, 1976, A Preliminary Report*, Geological Survey Paper 1002, U.S. Gov. Printing Office, Washington, DC.

Hoose, S.N., Wilson, R.C., and Rosenfeld, J.H. (1978). "Liquefaction-Caused Ground Failure During the February 4, 1976, Guatemala Earthquake," *Proceedings of the International Symposium on the February 4th Earthquake and the Reconstruction Process, 2.*

Krinitzsky, E.L. and Bonis, S.B. (1976). *Notes on Earthquake Shaking in Soils, Guatemala Earthquake of 4 February, 1976*, Informal Report, U.S. Army, Washington, DC.

Seed, H.B., Arango, I., Chan, C.K., Gomez-Masso, A., and Ascoli, R. (1981). "Earthquake-Induced Liquefaction Near Lake Amatitlan, Guatemala," *Journal of the Geotechnical Engineering*, 107(GT4), 501-518.

SCHOHARIE CREEK BRIDGE
(1987)

The Schoharie Creek Bridge was constructed in 1953 and carried the New York State throughway across the Schoharie Creek. It consisted of five simply supported spans with a total distance between the abutments of 165 m (540 ft) (Figure 1-2). The bridge carried two lanes of traffic in each direction approximately 24 m (80 ft) above the creek bed (Figure 1-3). On April 5, 1987, during the worst flooding experienced in years, the bridge collapsed (Figure 1-4). Four cars, a truck, and ten lives were lost as a result. The sudden rerouting of the highway disrupted commerce on both sides of the creek and focused attention on the collapse of an otherwise undistinguished structure.

Investigations confirmed that scour of the pier was the primary cause of failure. The primary defense of the Schoharie Creek Bridge against scour was the use of dry rip-rap. This consisted of a large field of quarry stones shaped as right rectangular prisms to inhibit rolling in flood waters. In early 1955 vertical cracks were observed in the pier plinths. A heavily reinforced concrete element 0.9 m (3 ft) thick was cast on top of each plinth in 1957 to control further cracking. Inspections were conducted in 1983 and 1986, but on the second occasion high water prevented detailed inspections of the bottoms of the piers.

During the April 1987 flood scour removed the support from the southern portion of the plinth, thereby generating tensile bending stresses in the top of the plinth leading to eventual fracture through the plinth and cross footing that served to connect the two support columns (Figure 1-5). An immediate loss of support between spans 3 and 4 was inevitable.

Lessons Learned

The failure due to scour of this bridge emphasizes the necessity for ensuring that bridge footings are deep enough to avoid the loss of support capacity arising from scour around the foundation. The presence of flood waters during the 1986 inspection inhibited a thorough inspection of the bridge. In hindsight, it appears to have been irresponsible for those involved not to have re-inspected when the flood had receded and it would have been possible to have undertaken a more comprehensive examination of the footings of the columns. The continuing problem with scour causing bridge collapses prompted interest in improved technology for underwater inspection.

It is important to accurately predict the effects of scour, and to design bridges to resist those effects. Lessons learned include:

1. Proper selection of a critical storm for the design of bridges crossing water.
2. The need for regular inspections of the superstructure, substructure, and underwater features of the bridge.

3. The importance of adequate erosion protection around piers and abutments susceptible to scour.

References

Civil Engineering (CE). (1988). "Lessons from Schoharie Creek," (1988), *CE*, 58(5), 46-49.

Delatte, Norbert J. (2009). *Beyond Failure: Forensic Case Studies for Civil Engineers*, ASCE Press, Reston, VA, 277-287.

Levy, M., and Salvadori, M. (1992). *Why Buildings Fall Down: How Structures Fail*, W. W. Norton, New York, NY, 143 – 147.

National Transportation Safety Board (NTSB). (1988). *Collapse of New York Thruway (1-90) Bridge over the Schoharie Creek, near Amsterdam, New York, April 5, 1987*, Highway Accident Report: NTSB/HAR-88/02, Washington, DC.

New York State Thruway Authority. (1987). *Collapse of Thruway Bridge at Schcubic yardsoharie Creek*, Final report, New York State Thruway Authority, November.

Storey, C., and Delatte, N. (2003). "Lessons from the Collapse of the Schoharie Creek Bridge," *Forensic Engineering: Proceedings of the Third Congress*, 158 – 167.

Swenson, D.V. and Ingraffea, A.R. (1991). "The Collapse of Schoharie Creek Bridge – A Case Study in Concrete Fracture Mechanics," *International Journal of Fracture*, 51, 73-92.

U.S. Senate Committee on Water Resources, Transportation and Infrastructure. (1987). *Collapse of the New York State Thruway Bridge over the Schoharie Creek*.

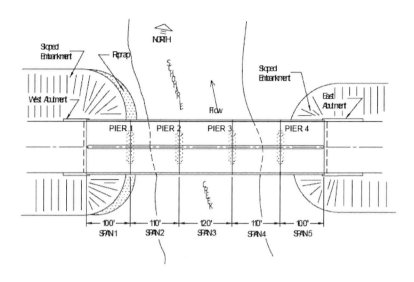

Figure 1-2. Schoharie Creek Bridge plan view.
Source: Storey and Delatte (2003), ©ASCE

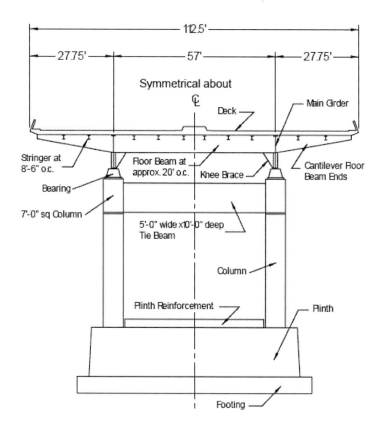

Figure 1-3. Schoharie Creek Bridge pier frames.

Source: Storey and Delatte (2003), ©ASCE

Figure 1-4. Schoharie Creek Bridge collapsed spans.
Source: Delatte (2009). Courtesy of Howard F. Greenspan (Howard F. Greenspan Associates).

Figure 1-5. Schoharie Creek Bridge failed plinth.
Source: Delatte (2009). Courtesy of Howard F. Greenspan (Howard F. Greenspan Associates).

Chapter 2

Embankment, Dam, and Slope Failures

ST. FRANCIS DAM
(1928)

On March 12, 1928 one of the worst civil engineering disasters of the 20th century occurred suddenly and with little warning. The St. Francis Dam, located 64 km (40 miles) northwest of Los Angeles, California, failed unexpectedly, releasing 47 million cubic meters (12.5 billion gallons) of water into San Francisquito Canyon. The result was devastating, taking the lives of over 400 people and leaving a wake of destruction.

The St. Francis Dam was designed and constructed by a self-taught and prominent engineer, William Mulholland. The Dam was initially designed as a stepped concrete gravity arch on a 152 m (500ft) radius, with an initial design height of 53 m (175 ft) from the floor of the canyon. The resulting reservoir capacity would be approximately 38 million cubic meters (30,000 acre-ft) of water. After several uncalculated design modifications to the height of the dam occurred during construction, the final dam height was 59 m (195 ft) with a reservoir capacity of 47 million cubic meters (38,168 acre-ft). Construction of the dam began in the spring of 1924 and concluded in 1926. However, the life of the dam was short as it failed only two years after completion.

Though a definitive cause for the failure was never determined, various theories have been proposed during the nearly 80 years since the disaster. With such theories as sabotage, geology, and poor construction, a variety of reasons for the failure have been investigated. It was concluded that whether or not the failure was entirely due to the geology of the site, the location in which the St. Francis Dam was situated was unsuitable for this type of structure. The varying geology between the two canyon walls and the inconsistent and unstable properties of the rock were unacceptable for a dam foundation.

Not only was the geology of the site in question, the design and construction techniques also came into view after the disaster. The failure to incorporate essential dam safety features into the design might have resulted in the collapse. A lack of uplift relief and expansion joints, along with drastic modifications of the dam during construction might have contributed to the demise of the St. Francis Dam. Without an outside source to verify the design as well as corroborate the design with the as-built structure, it is not clear as to whether the mistakes and decisions made by Mulholland would have been altered to provide stable dam design.

Lessons Learned

The lessons learned from the St. Francis Dam failure brought about several changes in the way future dams were to be designed. Extensive geological surveys of potential dam sites became an integral part of the design process. Uplift acting on the dam base became a major design consideration resulting in deeper foundations and seepage prevention. New laws were enacted by California that required any proposed dam design to be evaluated by an independent review panel before construction would be allowed.

References

Outland, Charles F. (2002). *Man-Made Disaster: The Story of St. Francis Dam*, Historical Society of Southern California, Los Angeles, CA.

Rogers, David J. (1995). *A Man, A Dam and A Disaster: Mulholland and the St. Francis Dam*, The Arthur H. Clark Company, Spokane, WA.

Shepherd, R. (2003). "The St. Francis Dam Failure," *Proceedings of the Third Forensic Engineering Congress*, ASCE Press, Reston, VA.

MALPASSET DAM
(1959)

The Malpasset Dam in Southern France was a double curvature arch dam with a maximum height of about 60 m (220 ft) and a crest length of about 223 m (730 ft). The thickness of the concrete varied from 1.5 m (5 ft) at the crest to 6.8 m (22 ft) at the center of the base. The dam created a reservoir with an estimated total capacity of about 51 million cu m (67 million cubic yards). The reservoir was filled very slowly over a five year period and at the time of failure the level of the water was 0.3 m (1 ft) below the spillway elevation.

Following a period of heavy rain, which resulted in the water level rising by almost 4 m (13 ft) in the three days immediately preceding the failure, it was decided to open the bottom outlet gate in the dam. This would permit a controlled release of water and prevent damage to a highway bridge under construction downstream of the dam. The response of the dam had been monitored intermittently during filling by survey measurements made on targets located on the downstream face of the dam.

Within hours of the bottom outlet gate being opened on December 2, 1959, the dam failed without warning. The city of Frejus, 7 km (4.5 miles) downstream of the dam suffered heavy losses resulting from the release of water and debris. More than 300 people died in the disaster. Blocks of concrete from the dam were washed as much as 1.5 km (1 mile) downstream.

The failure of Malpasset Dam represented the first failure of an arch dam. The suddenness of the failure, given that nothing abnormal had been detected at the dam within the hours preceding the event, added to the uncertainty.

Lessons Learned

A number of valuable lessons that changed design and construction methods for future dams resulted from the Malpasset Dam failure. Most notably, design and construction methods to mitigate the effects of uplift pressures in the foundation of the dam were developed. Further, the use of appropriately located and selected monitoring instruments increased following the dam failure.

References

Bellier, J. (1967). "Le Barrage de Malpasset," *Traveuse*, July.
Bellier, J. and Londe, P. (1976). "The Malpasset Dam," *Proceedings of Engineering Foundation Conference on the Evaluation of Dam Safety.*
Delatte, Norbert J. (2009). *Beyond Failure: Forensic Case Studies for Civil Engineers*, ASCE Press, Reston, VA, 267-277.
Engineering News-Record (ENR). (1959). "French Dam Collapse: Rock Shift was Probable Cause," *ENR*, December 10, 24-25.
Levy, M., and Salvadori, M. (1992). *Why Buildings Fall Down: How Structures Fail,* W. W. Norton, New York, NY.

Londe, P. (1987). "The Malpasset Dam Failure," *Engineering Geology*, 24, 295-329.

Ministere de l'Agriculture (1960). *Final Report of the Investigating Committee of the Malpasset Dam,* translated from the French and published by the U.S. Department of the Interior, National Science Foundation, and Israel Program for Scientific Translations, published in two volumes in 1963.

VAJONT DAM
(1963)

A 276 m (900 ft) high double-arched dam across the Vajont River valley in Northern Italy was constructed between 1957 and 1960. The dam created a reservoir with an estimated capacity of about 169 million cu m (220 million cubic yards). In 1959 concerns about potential slope stability problems in the reservoir were raised, which resulted in further analyses being undertaken. These studies confirmed that there was a slide problem; however, there was disagreement as to the volume of material which would be involved in a slide. Predictions ranged from relatively small volumes associated with local surficial movements 10 to 20 m (33 to 66 ft) deep up to volumes of the magnitude of the actual slide as a result of deep-seated movements.

Recognition of potential slope stability problems resulted in the installation of a monitoring program in 1960 and a staged reservoir filling schedule. It was considered that the rate of movements could be controlled by raising and lowering the reservoir water level. Results from the monitoring program over the next 3 years, which relied primarily on devices to detect surface movements, did confirm a relationship between the reservoir level and the slide mass movement; however, it failed to give a warning of the rate at which the failure ultimately occurred.

On October 9, 1963, the southern rock slope of the reservoir failed over an approximately 2 km (1.2 miles) length. Movement rates of the slide mass of approximately 275 million cu m (360 million cubic yards) during the failure were estimated to be of the order of 25 m per sec (80 ft per sec) as opposed to the typical rates recorded during the previous three years of monitoring of less than 1 cm per day (0.4 in. per day) to at most 20 cm per day (8 in. per day) on the day of the failure. The massive slide mass came to rest approximately 360 m (1,180 ft) laterally, and 140 m (429 ft) upward, on the opposite bank of the reservoir (Figure 2-1). The top of the slide mass was 160 m (525 ft) above the crest of the arch dam. At the time of the failure, the reservoir was about 66% full and contained an estimated 115 million cu m (150 million cubic yards.) of water. The water level had been lowered in a controlled fashion by approximately 10 m (33 ft) in the preceding two weeks. As the slide mass plunged into the reservoir, the water was displaced over the dam crest in a stream estimated to be up to 245 m (800 ft) above crest level. Five villages and 2,040 lives were lost.

Lessons Learned

The massive slide into the Vajont reservoir yielded important lessons in regard to the analysis and monitoring of slope movements. The difficulty of predicting when a slide mass will accelerate or fail became evident and the difficulty of estimating changes in states of stress and strength during sliding was reinforced.

References

Delatte, Norbert J. (2009). *Beyond Failure: Forensic Case Studies for Civil Engineers*, ASCE Press, Reston, VA, 234-248.

Hendron, A.J., and Patton, F.D. (1985). *The Vajont Slide: A Geotechnical Analysis Based on New Geological Observations of the Failure Surface*, Volume I, Main Text, Technical Report GL-85-5, Department of the Army, U.S. Army Corps of Engineers, U.S. Army Engineer Waterways Experiment Station, Vicksburg, MS.

Kiersch, G.A. (1964). "Vajont Reservoir Disaster," *Civil Engineering*, 32-40.

Muller, L. (1964). "The Rock Slide in the Vajont Valley," *Rock Mechanics and Engineering Geology*, II, 148-212.

Muller, L. (1968). "New Considerations in the Vajont Slide," *Rock Mechanics and Engineering Geology*, VI, 1-91.

Muller, L. (1987). "The Vajont Catastrophe - A Personal Review," *Engineering Geology*, 24, 423-444.

Wearne, Phillip (2000). *Collapse: When Buildings Fall Down*, TV Books, L.L.C. (www.tvbooks.com), New York, NY.

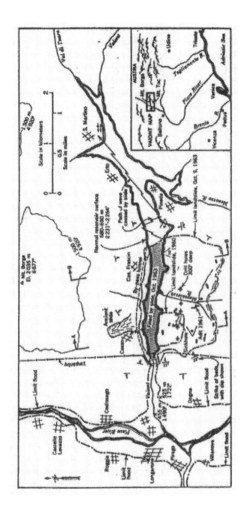

Figure 2-1. Map of Vajont Dam Area Showing Limits of Slide and Resulting Wave.
Source: Kiersch (1964), ©ASCE

LOWER SAN FERNANDO DAM
(1971)

Construction of a 40 m (130 ft) high hydraulic-fill earth dam was initiated in 1912 as part of a reservoir system in San Fernando, California. Hydraulic fill was placed between 1912 and 1915. The material was excavated from the bottom of the reservoir and discharged through sluice pipes located at starter dikes on the upstream and downstream edges of the dam. This construction configuration resulted in upstream and downstream shells of sands and silts and a central core region of silty clays. A stratum of variable thickness between 3 and 5 m (10 and 16 ft) of reworked weathered shale (silty sand to sand size) was placed on top of the hydraulic fill material in 1916. Several additional layers of roller compacted fill were placed between 1916 and 1930 and raised the dam to its final height of about 40 m (130 ft). A roller compacted berm was placed on the downstream side in 1940. The dam created a reservoir with an estimated full capacity of about 25 million cu m (33 million cubic yards).

On February 9, 1971, the San Fernando earthquake occurred with an estimated Richter magnitude of 6.6. At the time of the earthquake the water level in the reservoir was about 11 m (36 ft) below the crest. This reduced level was in part the result of an earlier seismic stability analysis that imposed a minimum operating freeboard criterion of 6 m (20 ft). During and immediately after the earthquake, a major slide involving the upstream slope and the upper part of the downstream slope occurred (Figure 2-2). As a result of the slide a freeboard of about 1.5 m (5 ft) remained. Given the likelihood of further damage in the presence of after-shocks, 80,000 people living downstream of the dam were evacuated over a 4-day period until the water level was lowered to a safe elevation.

Lessons Learned

The near-disastrous slide of the San Fernando dam had a major impact on future earth dam design and construction procedures. The availability of site specific seismograph data permitted extensive analyses to be performed and showed among other factors, the potential problems with hydraulic fill structures and the need for revised procedures for dynamic stability analyses of earth dams.

References

Castro, G., Seed, R.B., Keller, T.O., and Seed, H.B. (1992). "Steady State Strength Analysis of Lower San Fernando Dam Slide," *Journal of Geotechnical Engineering Division*, 118 (GT3), 406-427.

Seed, H.B., Idriss, I.M., Lee, K.L., and Makdisi, F.I. (1975). "Dynamic Analysis of the Slide in the Lower San Fernando Dam during the Earthquake of February 9, 1971," *Journal of Geotechnical Engineering Division*, 101 (GT9), 889-911.

Seed, H.B., Lee, K.L., Idriss, I.M., and Makdisi, F.I. (1975). "The Slides in the San Fernando Dams during the Earthquake of February 9, 1971," *Journal of Geotechnical Engineering Division*, 101 (GT7), 651-688.

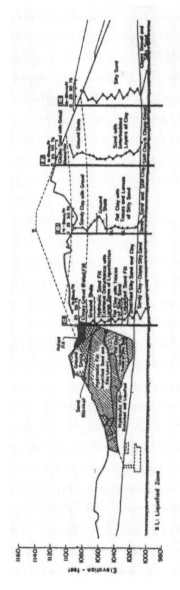

Figure 2-2. Post-Earthquake Section through Lower San Fernando Dam.
Source: Seed, et al. (1975), ©ASCE

TETON DAM
(1976)

A 90 m (300 ft) high zone-filled earth dam was constructed in a steep walled canyon eroded by the Teton River in Idaho (Figure 2-3). The typical cross-section (Figure 2-4), shows a wide silt core, with upstream and downstream shells consisting mainly of sand, gravel and cobbles. In the main section of the dam, the impervious core was keyed into the foundation alluvium 30 m (100 ft) deep to serve as a cut-off trench. Lesser cut-off trenches were excavated at both abutments through the permeable rock. Reservoir filling commenced in November 1975 at an intended rate of about 0.3 m (1 ft) per day. Delays in completing the construction of outlet works combined with heavier than expected spring melt run-off resulted in a filling rate up to 1.2 m (4 ft) per day in May 1976.

The dam failed on June 5, 1976, when the water level in the reservoir was at an elevation 9 m (30 ft) below the embankment crest and 1 m (3 ft) below the spillway crest. Breaching of the dam crest and complete failure was preceded over a period of two days by increasing quantities of seepage. This seepage was observed initially 460 m (1500 ft) downstream and later on the downstream face of the dam (Figure 2-5). Noticeable increase in seepage rate from the face of the dam adjacent to the abutment about 40 m (130 ft) below the crest occurred during the morning of June 5. By approximately 10:30 am, the flow rate of seepage increased to about 0.4 cu m per sec (15 cu ft per sec). This quantity continued to increase as a 1.8 m (6 ft) diameter "tunnel" formed perpendicular to the longitudinal axis of the dam. By 11:00 am a vortex was observed in the reservoir. The seepage flow rate increased rapidly from this time onwards, accompanied by progressive upward erosion of the "tunnel" crown (Figures 2-6 through 2-9). The dam crest was breached at about 11:55 am, with complete failure of the dam ensuing. The flooding downstream after the failure of the dam resulted in the loss of 14 lives and caused an estimated $400 million in damage.

Lessons Learned

The failure of Teton Dam was important in that it was the tallest dam to have failed. It provided important lessons relating to the need for instrumentation; the need for protective filters to prevent uncontrolled seepage erosion; the design of cut-off trenches; consideration of the impact of frost action; and the importance of adequate compaction control criteria and methods.

References

Chadwick, W.L. (1977). "Case Study of Teton Dam and its Failure," *Proceedings of 9th International Conference on Soil Mechanics and Foundation Engineering*, Case History Volume.

Delatte, Norbert J. (2009), *Beyond Failure: Forensic Case Studies for Civil Engineers*, ASCE Press, Reston, VA, 223-234.

Independent Panel to Review Cause of Teton Dam Failure (1976). *Report to the U.S. Department of the Interior and State of Idaho on Failure of Teton Dam,* Idaho Falls, ID.

Leonards, G.A., and Davidson, L.W. (1984). "Reconsideration of Failure Initiating Mechanisms for Teton Dam," *Proceedings of 1st International Conference on Case Histories in Geotechnical Engineering,* 3, 1103-1113.

Seed, H.B., and Duncan, J.M. (1981). "The Teton Dam Failure – A Retrospective Review," *Proceedings of 10th International Conference on Soil Mechanics and Foundation Engineering,* 4, 219-238.

Seed, H.B., and Duncan, J.M. (1987). "The Failure of Teton Dam," *Engineering Geology,* 24, 173-205.

Solava, S., and Delatte, N. (2003). "Lessons from the Failure of the Teton Dam," *Proceedings of the Third Forensic Engineering Congress,* 168 – 177.

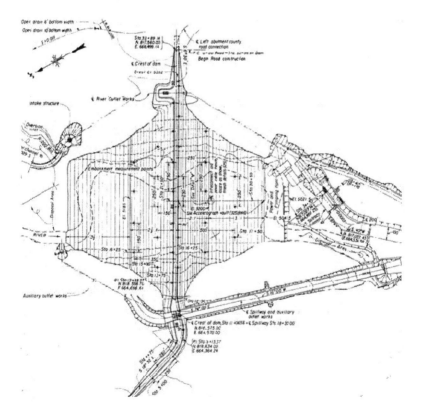

Figure 2-3. Teton Dam Construction Details, Plan View.
Source: Independent Panel (1976).

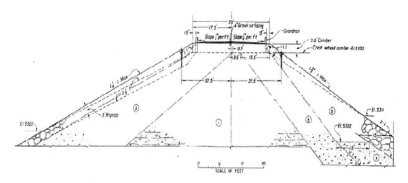

Embankment Explanation
1. Selected clay, silt, sand, gravel, and cobbles compacted by tamping rollers to 150 mm (6-inch) layers.
2. Selected sand, gravel, and cobbles compacted by crawler-type tractors to 300 mm (12-inch) layers.
3. Miscellaneous material compacted by rubber-tired rollers to 300 mm (12-inch) layers.
4. Selected silt, sand, gravel, and cobbles compacted by rubber-tired rollers to 300 mm (12-inch) layers.
5. Rockfill placed in 0.9 m (3-foot) layers.

Figure 2-4. Teton Dam Construction Details, Cross Section.
Source: Independent Panel (1976).

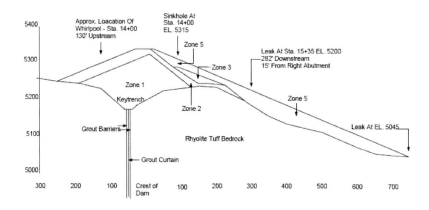

Figure 2-5. Teton Dam Initial Failure Indications.
Source: Independent Panel (1976).

Figure 2-6. Teton Dam Failure Sequence, Initial Erosion.
Source: Independent Panel (1976).

Figure 2-7. Teton Dam Failure Sequence, the Seepage Widens.
Source: Independent Panel (1976).

Figure 2-8. Teton Dam Failure Sequence, Beginning of Final Failure.
Source: Independent Panel (1976).

Figure 2-9. Teton Dam Failure Sequence, Failure of the Dam.
Source: Independent Panel (1976).

RISSA NORWAY LANDSLIDE
(1978)

A landslide overran seven farms in an area of 330,000 sq m (0.127 sq miles) in Rissa, Norway on April 29, 1978. Rissa is located just north of the city of Trondheim, Central Norway. The Trondheim and Oslo regions are the two large marine clay areas of Norway. The Rissa event involved 5 to 6 million cu m (7 to 8 million cubic yards) of slide debris. It is the biggest landslide to have occurred in Norway in the 20th century. Of forty people caught within the slide area only one was killed. Many people witnessed the event, which was also recorded on amateur films. The slide is well documented.

The landslide was due to complete liquefaction of quick clay and took place retrogressively in large sections triggered by the excavation and stockpiling of 700 cu m (900 cubic yards) of soil placed by the shore of Lake Botnen. Excavation work for the construction of a new wing to an existing dam was conducted over a two day period. The surplus soil had been placed down by the lake shoreline to extend farm area. When the earthwork was completed, 70 to 90 m (230 to 295 ft) of the shoreline suddenly slid into the lake. The sliding then developed retrogressively landward from that point. The total duration of the retrogressive sliding occurred over a period of about 6 minutes.

Lessons Learned

Construction in quick clay areas must be preceded by careful study and analysis.

References

Grande, L. (1978). *Description of the Rissa Landslide, 29 April 1978, Based on Eyewitness Accounts*, Tech. Univ. of Norway, Trondheim, Institute of Soil Mechanics and Foundation Engineering.
Gregersen, 0. (1981). *The Quick Clay Landslide in Rissa, Norway. The Sliding Process and Discussion of Failure Modes*, NGI Publication No. 135, Norwegian Geotechnical Institute, 1-6.

NERLERK BERM FAILURE
(1983)

Construction of an underwater sand berm designed as part of an offshore artificial sand island structure for hydrocarbon exploration at Nerlerk in the Canadian Beaufort Sea was undertaken during the 1982 and 1983 open-water construction seasons. A series of slides occurred during the 1983 construction season that ultimately led to abandonment of the berm.

The artificial island was to be constructed at a site where the water was about 45 m (150 ft) deep. The design called for a sand berm with side slopes of the order of 5H:1V extending from the seafloor to the design berm top elevation of about 10 m (33 ft) below sea level. A steel caisson superstructure would then be ballasted on top of the berm to permit hydrocarbon exploration. About 1.3 million cu m (1.7 million cubic yards) of dredge material was placed during the 1982 construction season using bottom dump hopper dredges, which hauled the medium sand fill from the Ukalerk borrow area. An additional 1.8 million cu m (2.3 million cubic yards) of clean medium sand material was dredged during the 1982 construction season from a nearby borrow area and was pumped via floating pipeline to a discharge barge situated at the berm site.

Construction recommenced in the summer of 1983 with the deep suction dredger to discharge to a barge floating pipeline system. Bathymetric surveys revealed that a significant part of the berm disappeared by July 20, 1983. Several additional construction slides were noted in the following weeks as efforts were made to complete the berm to the design elevation and assess the cause of the initial failure. Construction at the initial site was finally abandoned in early August. Hydrocarbon exploration activities were ultimately conducted using an alternative drilling system. Nevertheless, berm construction losses in excess of $50 million were incurred.

Lessons Learned

This case history showed the need to carefully control dredge construction methods. Post failure investigations indicated fill materials in certain zones were in a very loose state.

References

Been, K., Conlin, R.H., Crooks, J., Fitzpatrick, S.W., Jefferies, M.G., Rogers, R.T., and Shinde, S. (1987). Discussion of "Back Analysis of the Nerlerk Berm Liquefaction Slides," *Canadian Geotechnical Journal*, 24 (1), 170-179.

Konrad, J.M. (1991). "The Nerlerk Berm Case History: Some Considerations for the Design of Hydraulic Sand Fills," *Canadian Geotechnical Journal*, 28 (4), 601-612.

Sladen, J.A., D'Hollander, R.D., Krahn, J., and Mitchell, D.E. (1985). "Back Analysis of the Nerlerk Berm Liquefaction Slides," *Canadian Geotechnical Journal*, 22 (4), 579-588.

Sladen, J.A., D'Hollander, R.D., Krahn, J., and Mitchell, D.E. (1987). Closure to "Back Analysis of the Nerlerk Berm Liquefaction Slides," *Canadian Geotechnical Journal*, 24 (1), 179- 185.

CARSINGTON EMBANKMENT
(1984)

A 1250 m (410 ft) long, 30 m (100 ft) high zone-filled earth embankment was being constructed as part of a water storage reservoir for the Severn-Trent Water Authority to regulate flows in the River Derwent in England. The reservoir was designed so that water would be diverted through a 10 km (6 mile) long tunnel during the winter and stored in the reservoir with an estimated capacity of 35 million cu m (46 million cubic yards). Water would be released from the reservoir when the water level in the river was low.

The central clay core of the embankment connected to a shallow trench that was excavated upstream of the centerline into the weathered gray foundation mud stone. A grout curtain extended below the base of the trench. Fill in the upstream and downstream shells was classified as Type I and Type II. The Type I fill, used immediately upstream and downstream of the clay core, was described as a yellow-brown mottled clay with mud stone peas < 5 mm (0.2 in.) and pebbles. The Type II soil specified for the outer portions of the shells was to be the same general type of material as Type I, but without pebbles.

Construction of the embankment began in July, 1982 and reached a height of about 6 m (20 ft) at the end of the construction in late October. Construction during the second year between April and October added an additional 15 m (50 ft) to the embankment. Placement of an additional 4 m (13 ft) of fill took place in the two months preceding the failure. Tension cracks over an approximate 65 m (213 ft) length were first observed on the crest of the embankment on June 4, 1984 when about 1 m (3 ft) of crest remained to be placed. Thirty six hours later, a 400 m (1,300 ft) long section of the upstream slope failed with a maximum horizontal displacement of 15 m (50 ft).

Lessons Learned

The failure of Carsington Embankment is considered of importance in that it led to additional attention being given to the role of the construction equipment and procedures in the subsequent stability of a structure. In this case, the compaction equipment selected and the rate of fill placement are considered to have been key factors in the observed failure. In addition, the importance of selecting instrumentation, which can provide a precursor to a failure, was reinforced.

References

New Civil Engineer (1984). "Weak Ground Cited as Carsington Fails," June 14.
Rowe, P.W. (1991). "A Reassessment of the Causes of the Carsington Embankment Failure," *Geotechnique*, 41 (3), 395-421.
Skempton, A.W. (1985). "Geotechnical Aspects of the Carsington Dam Failure," *Proceedings of 11th International Conference on Soil Mechanics and Foundation Engineering*, 5, 2581-2591.

Chapter 3

Geoenvironmental Failures

LOVE CANAL
(1978)

In the late 1890s, an entrepreneur named William T. Love initiated the construction of a canal linked to the Niagara River that was intended to be used as a source for hydroelectric power and attract industry to the Niagara Falls, New York area. The project was abandoned when a section of canal about 1000 m (3,200 ft) long and 24 m (80 ft) wide had been excavated to a depth of 6 m (20 ft). In 1942, the Hooker Chemical and Plastic Company purchased the abandoned excavation site from the Niagara Power and Development Company and began using the canal excavation as a dumpsite for industrial wastes that included pesticide residues, process slurries, and waste solvents. In total, approximately 20,000 tonnes (22,000 tons) of waste contained in metal drums was placed in the excavation during an eleven year period. Once filled, the excavation was capped with a loose soil cover.

The Hooker Chemical and Plastic Company sold the land to the City of Niagara Falls for $1 in 1953. A residential sub-division and school were subsequently built on the site. During the mid -1970s area residents began developing a variety of illnesses. In 1975 and 1976 significant precipitation raised the groundwater table and caused portions of the land to subside. Waste drums appeared at the surface and chemical odors were very noticeable near sewer manholes. Some of the houses nearest the old canal site had basements where black sludge that produced a strong smell was observed oozing through the walls. After several preliminary studies revealed increased rates of birth defects and miscarriages among area residents, the government declared a state of emergency and two separate evacuations took place in 1978 (area immediately adjacent to canal involving approximately 240 families) and in 1980 a larger area involving approximately an additional 570 families). Later studies of the soil and industrial wastes would show that more than 200 different chemical compounds including at least 12 known carcinogens were present. Extensive and protracted clean-up at the site including removal of about 5,700 cu m (7,500 cubic yards) of contaminated soil, placement of a 1-m (3 ft) clay cover over 7 ha (18 acres) immediately above and adjacent to the canal, placement of a second composite high density poly ethylene (HDPE)clay liner system over 16 ha (40 acres) and cleaning of about 200 m (5,000 ft) of sewer resulted in estimated costs in excess of half a billion dollars to date and there remains debate about the future habitability of the area.

Lessons Learned

The Love Canal site was one of the landmark environmental failures that came to light in the 1970s and led to the passing of the Comprehensive Environmental Response, Compensation and Liability Act EPA, Superfund Record of 1980. As such, it changed the attention given to environmental assessment prior to real estate transactions and the assignment of responsibility for inappropriate disposal of hazardous wastes.

References

Brown, Michael H. (1979), "Love Canal and the Poisoning of America," The Atlantic, December, 33-47.

Colten, C. and Skinner, P. (1996). The Road to Love Canal: Managing Industrial Waste Before EPA, University of Texas Press, Austin, TX.

DeLaney, K. (2000). "The Legacy of the Love Canal in New York State's Environmental History". *Proc., Researching New York: Perspectives on Empire State History Annual Conf.*

Fletcher, T. (2001). "Neighborhood Change at Love Canal: Contamination, Evacuation, and Resettlement," *Land Use Policy*, 19, 311-323.

Gibbs, L. M. (1981). *Love Canal: The Story Continues*, New Society Publishers, Gabriola Island, British Columbia, (http://onlineethics.org/edu/precol/classroom/cs6.html) (Sept. 21, 2005); (http://en.wikipedia.org/wiki/Love-Canal) (Sept. 21, 2005); (http://ublib.buffalo.edu/libraries/projects/lovecanal/science-gif/records/hart1.html) (Sept. 21, 2005); (http://www.epa.gov/region2/superfund/npl1/0201290c.pdf) (Sept. 21, 2005).

Levine. A. G. (1982). *Love Canal: Science, Politics, and People*, Lexington Books, Lexington, MA.

Love Canal Collection; http://library.buffalo.edu/libraries/specialcollections/lovecanal/index.html.

The Love Canal Tragedy; http://www.epa.gov/history/topics/lovecanal/01.htm.

New York State Department of Health. (1978). "Love Canal: A Special Report to the Governor and Legislature," State of New York, Albany, NY

New York State Department of Transportation Files. (1989). "NYSDOT Files of Commissioner William Hennessey," *Rep. 13430-89*, Love Canal Task Force and Office of Legal Affairs, New York State Archives.

New York State Task Force on Toxic Substances Files (1981). "New York State Task Force on Toxic Substances Files of Chairman and Assemblyman Maurice Hinchley," *Rep. 1.0134*, New York State Archives.

Phillips, A. S., Hung, Y., and Bosela, P. (2007), "Love Canal Tragedy," *Journal of Performance of Constructed Facilities* 21(4), 313-319.

Silverman, G. (1989). "Love Canal: A retrospective," *Environmental Reporter*, 20(20-2), 835-850.

Superfund: History of Failure; http://www.ncpa.org/pdfs/ba198.pdf.

TIME (1988), "Welcome Back to Love Canal," October 10, Vol. 132, 49.

U.S. EPA, Environmental Monitoring at Love Canal (1982), EPA 600/4-82-0309.S. Decision: Love Canal, 93rd Street (1988), NY, September, EPA RODIR02-88/063.

VALLEY OF THE DRUMS
(1978)

The Valley of the Drums is located in Northern Bullitt County, Kentucky near the town of Brooks on land that was owned by Mr. A.L. Taylor prior to his death in 1977. The site, which was first identified as a waste disposal facility by the Kentucky Department of Natural Resources and Environmental Protection (KDNREP) in 1967, consisted of approximately 5 ha (13 acres) that was used as an uncontrolled dump site.

More than 27,000 drums of industrial waste were discovered on the site in 1978 by KDNREP officials. Later studies however, estimated that more than 100,000 drums had been delivered to the site. Due to space limitations, the contents of many of the drums were dumped in open pits and trenches that were excavated on site. The empty drums were then either sold or crushed. The trenches were subsequently covered with on-site soil fill.

Over a period of time, the conditions of many of the drums on site deteriorated and the contents spilled onto the ground and were flushed into a nearby creek by storm water runoff. Additional contaminants seeped into the creek from the disposal trenches. Frequent complaints about strong odors along the creek bed were received from adjacent property owners. In March 1979 unusually large amounts of wastes were flushed into the creek by melting snow runoff and resulted in an emergency response by the Environmental Protection Agency. Subsequent analysis of water and soil samples indicated substantial levels of contamination by heavy metals and polychlorinated biphenyls along with 140 other chemical substances. Extensive remedial actions were undertaken in 1986 and 1987 to control surface run-off and reduce the future impact of buried wastes.

Lessons Learned

The Valley of the Drums site was one of the landmark environmental failures that came to light in the 1970s and led to the passing of the Comprehensive Environmental Response, Compensation and Liability Act in 1980. As such, it changed the attention given to facility permitting and compliance monitoring and the assignment of responsibility for inappropriate disposal of hazardous wastes.

References

Boving, T.B. and Blue, J. (2002). "Long-Term Contaminant Trends at the Picillo Farm Superfund Site in Rhode Island," *Remediation Journal*, 12 (2) 117-128.

Hale, D. R. (1998). "Buying Time: Franchising Hazardous and Nuclear Waste Cleanup," *Fuel and Energy Abstracts*, 39(1), 30.

Hamilton, V., and W. K. (1999). "How Costly is "Clean"? An Analysis of the Benefits and Costs of Superfund Site Remediations," *Journal of Policy Analysis and Management*, 18 (1), 7

Metcalf & Eddy, Inc. (1984). *Feasibility Study Addendum and Endangerment Assessment: A.L. Taylor Site*, EPA Contract #68-01-6769, September.

Resource Applications, Inc. (1992). *A.L. Taylor Five-Year Review Final Report*, EPA Contract #68- W9-0029, June.

US Environmental Protection Agency (USEPA) Oil and Special Materials Control Division (1980). *Valley of the Drums, Bullitt County, Kentucky, August 1980*, Publication No. EPA-430/9-80-014, USEPA, Washington, D.C.

USEPA. (1986*). Record of Decision, Remedial Alternative Selection, A.L. Taylor Site, Brooks, Kentucky*. Region IV Administrative Records Division, Atlanta, GA., Document No. 001422, 109 .

STRINGFELLOW ACID PITS
(1980)

Between 1956 and 1972 the Stringfellow Quarry Company operated a state authorized hazardous waste disposal facility 8 km (5 miles) Northwest of Riverside, California. Approximately 155 million liters (34 million gallons) of industrial wastes from metal finishing, electroplating and Department of Transportation (DOT) production industries were placed in unlined evaporation ponds. The wastes disposed of in these ponds migrated into the underlying highly permeable soils into the groundwater and resulted in a contaminated plume which extended some 3 km (2 miles) downstream.

The site was voluntarily closed in 1972. The California Regional Water Quality Control Board declared the site a problem area. Between 1975 and 1980, approximately 30 million liters (6.5 million gallons) of liquid wastes and DOT contaminated materials were removed and a policy to contain the waste and minimize further contaminant migration was adopted. The Environmental Protection Agency (EPA) led an additional clean-up effort in 1980 that resulted in an additional 45 million liters (10 million gallons) of contaminated water being removed from the site. In 1983, the site was added to the National Priorities List as California's worst environmental hazard. Since that time, 4 Records of Decision have been issued by the EPA outlining required clean-up measures. Estimates of clean-up costs as high as three-quarters of a billion dollars have been cited along with indications that the work could last for decades.

Lessons Learned

The Stringfellow Acid Pits site was one of the landmark environmental failures that came to light in the 1970s and led to the passing of the Comprehensive Environmental Response, Compensation and Liability Act in 1980. As such, it changed the attention given to environmental assessment prior to real estate transactions and the assignment of responsibility for inappropriate disposal of hazardous wastes. One interesting fact about this site is that it is one of the few cases where a government has been found liable for contributing to an environmental problem. In this case, the State of California was found at fault as a result of its actions in selecting this particular site for disposal and subsequently controlling activities there.

References

Engineering News Record (ENR) (1992). "EPA, PRP's Sign Pact to Clean Stringfellow," *ENR*, August 17, 229, 14.
New York Times (NYT) (1993). "Largest-ever Toxic-Waste Suit Opens in California," *NYT*, February 5, 142.

USEPA (1983). *Stringfellow Record of Decision: Stringfellow Acid Pits Site; Initial Remediation Measure*, EPA/ROD/R09-83/005, USEPA, July.

USEPA (1984). *Stringfellow Record of Decision: Stringfellow Acid Pits: Untitled*, EPA/ROD/R09-84/007, USEPA, July.

USEPA (1987). *Stringfellow Record of Decision: Stringfellow Hazardous Waste Site: 2nd Remedial Action*, EPA/ROD/R09-87/016, USEPA, July.

USEPA (1990). *Stringfellow Record of Decision: Stringfellow Hazardous Waste Site: 4th Remedial Action*, EPA/ROD-R09-90/048.

SEYMOUR RECYCLING FACILITY
(1980)

The Seymour Recycling Corporation operated a 5.6 ha (14 acre) site as a recycling facility to process industrial waste chemicals approximately 3 km (2 miles) southwest of the city of Seymour in Jackson County, Indiana. From 1970 to 1980, wastes were collected in drums and bulk storage tanks. By 1980, there were approximately 100 storage tanks and about 50,000 drums on site. Many of the drums were in poor condition and their contents had leaked while others had no lids. Widespread contamination of the underlying soil and groundwater occurred and resulted in on-site fires and unpleasant toxic odors being reported by neighboring residents.

As a result of a fire in 1980 chemical runoff from the site posed a threat that resulted in approximately 300 people being temporarily evacuated. The facility was subsequently closed. Over the next 4 years, the majority of tanks and drums were removed by those identified as being potentially the responsible parties.

A range of remedial measures were implemented. An embankment was constructed around the site to control surface runoff. Approximately 900,000 liters (200,000 gallons) of flammable chemicals were incinerated. More than 450,000 liters (100,000 gallons) of inert liquids were injected into a deep well. Other wastes and containers including more than 23,000 cu m (30,000 cubic yards) of drums, sludge and contaminated soil were placed in a hazardous waste landfill. Despite these remedial measures, monitoring wells showed that a contaminated groundwater plume extended more than 120 m (400 ft) off site by 1985. Tests indicated the presence of heavy metals and numerous organic compounds and phenols within the soil and groundwater. The site was placed on the National Priorities List as a result of being designated the most serious environmental threat in Indiana. Extensive long term remedial activities included on-site incineration of some contaminated soils, use of bioremediation technology to assist in the cleanup, the installation of a vapor extraction system to remove volatile organic compounds from the vadose zone and the installation of a pump and treat system to stabilize the contaminated groundwater plume and treat it at the Seymour waste water treatment plant.

Lessons Learned

The Seymour Recycling Facility site was one of the landmark environmental failures that occurred in the 1970s and contributed to the passing of the Comprehensive Environmental Response, Compensation and Liability Act in 1980. As such, it changed the attention given to facility permitting and compliance monitoring and the assignment of responsibility for inappropriate disposal of hazardous wastes.

References

USEPA (1984). *Hazardous Waste Sites, Description of Sites on Current National Priorities List, October 1984*, EPA/HW 8.5, USEPA, December.

USEPA (1986). *Superfund Record of Decision: Seymour, IN*, EPNRODIR0586/046, USEPA, September.

USEPA (1987). *Superfund Record of Division: Seymour, IN*, EPNRODIR0587/050, USEPA, September.

KETTLEMAN HILLS WASTE LANDFILL
(1988)

Prior to 1987, construction of a 15 ha (36 acre) landfill known as Unit B-19, began with the excavation of a 30 m (100 ft) deep oval-shaped "bowl" for a Class I hazardous-waste treatment-and-storage facility at Kettleman City, California. The "bowl" was designed to have a nearly horizontal base with steep side slopes into which waste fill was to be placed. A multilayer liner system consisting of impervious geomembranes, clay layers, and drainage layers, lined the base and sides of the repository to prevent the escape of hazardous materials (leachates) into the underlying and surrounding ground and underlying ground water (Figure 3- 1).

The lining of 6 ha (15 acres) (Phase I-A) of the northern end of the "bowl" was completed first and placement of solid hazardous waste began in early 1987. On March 19, 1988, after the waste pile reached a maximum height of about 27 m (90 ft) in Phase I-A with no prior indication of distress, a slope stability failure occurred with lateral displacements of the waste fill of up to 11 m (35 ft) and vertical settlement of the surface of the fill of up to 4 m (14 ft). Surface cracks, tears and displacements of the exposed portions of the liner system were also visible. It was found that failure developed by sliding along interfaces within the composite, multilayered geosynthetic compacted clay liner system beneath the waste fill.

Because of fear that the liner system may have also been breached, and to preclude the possibility of similar failures at other facilities, an investigation was undertaken to determine (1) the cause of the failure and (2) appropriate methods of testing and analysis. Comprehensive laboratory tests were performed to evaluate the frictional resistance between contact surfaces of various geosynthetics, geonets, and geotextiles; and between these materials and compacted clay liner. The frictional resistance at the interface between these materials was characterized as being insufficient to maintain stability of the waste pile.

Lessons Learned

The interface frictional resistance of liner materials is affected by various properties. These properties include degree of polishing, whether the interfaces are wet or dry, and in some cases, the relative orientation of the layers to the direction of shear stress application. Some small variations in properties also exist between one batch and another of the geosynthetic materials. Wetting of the compacted clay liner at its contact with the geosynthetics also affects the interface frictional resistance. The test and observed field results also indicate the desirability of performing similar test programs for proposed new facilities to establish design parameters, until such a time as more data and experience are available.

References

Martin, J.P., Koerner, R.M. and Whitty, J.E. (1984). "Experimental Friction Evaluation of Slippage between Geomembranes, Geotextiles and Soils," *Proceedings of International Conference on Geomembranes, Denver, CO*, 191-196.

Mitchell, J.K., Seed, R.B. and Chang, M. (1993). "The Kettleman Hills Landfill Failure: A Retrospective View of the Failure Investigations and Lessons Learned," *Proc. Third International Conference on Case Histories in Geotechnical Engineering, St. Louis, MO* Vol. II, 1379-1392.

Mitchell, J. K., Seed, R.B. and Seed, H.B. (1990). "Kettleman Hills Waste Landfill Slope Failure. I: Liner-System Properties," *Journal of Geotechnical Engineering*, 116(4) 647-668.

Seed, R.B., Mitchell, J.K. and Seed, H.B. (1990). "Kettleman Hills Waste Landfill Slope Failure. II: Stability Analysis," *Journal of Geotechnical Engineering*, 116(4) 669-690.

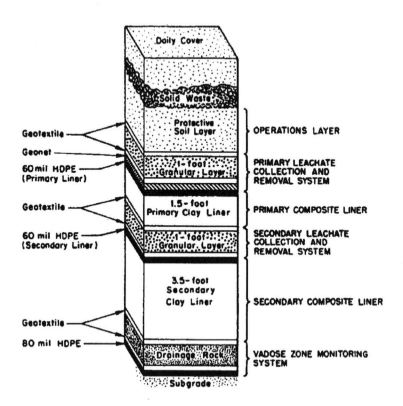

Figure 3-1. Schematic Illustration of Multilayer Liner System at Base of Kettleman Hills Landfill.
Source: Seed, et al. (1990), ©ASCE

NORTH BATTLEFORD, SASKATCHEWAN WATER TREATMENT FAILURE (Cryptosporidium Outbreak)
(2001)

In the spring of 2001, an outbreak of Cryptosporidiosis occurred in the town of North Battleford, Saskatchewan, causing approximately 6,500 people to become ill. *Cryptosporidium parvum* is a gastro-intestinal parasite transported by an oocyst, which is resistant to traditional water treatment. The oocyst leaves the parent host with fecal material, is ingested by mouth (through poor sanitation, such as dirty hands or contaminated water) and enters the digestive system of the new host. Symptoms of the disease are diarrhea, abdominal cramps, vomiting, and headaches.

Failure of the city's surface water treatment plant, located downstream from their water treatment plant, was found to be the source of the outbreak (Figure 3-2). The Sedimentation Contact Unit (SCU), which had been re-started after the plant had been shut down for planned maintenance, did not satisfactorily remove suspended solids from the source water, and contaminated water was released into the distribution system. Communication breakdown and confusion, triggered by personnel turnover, insufficient budget and lack of effective communication with the Saskatchewan Environment and Resource Management (SERM) resulted in an extended exposure to the public.

Physical changes were made in the water treatment system, including new valves and pipes added at the SCU and filter banks so that water could be run to waste (Figure 3-3). Sampling sites were added after the filter runs to allow for manual water collection and continuous monitoring of turbidity and particulates. Coagulant chemicals were added at the sand separators for better mixing. Ultraviolet (UV) disinfection units were added after both filter banks. UV light has been shown to be more effective than chemicals for inactivating *Cryptosporidium* oocysts. Chlorine disinfection was moved to the clearwells instead of before the SCU. Chlorination units were also installed for the town's water supply reservoirs within the distribution system. Chlorine disinfection was increased at the sewage plant for the purpose of treating bypass sewage.

Finally, Standard Operating Procedures (SOPs) were added for the purposes of quality assurance. The new SOPs called for the surface water plant to shut down or run water to waste when the following occur:

1. Sewage bypass events
2. Chlorination malfunction
3. Malfunction in the SCU
4. Finished water with a turbidity greater than 0.3 NTU

Some of the additional changes included:

1. Increased funding for the regulatory branch of SERM as well as more personnel.

2. At least one inspection of drinking water and sewage plants per year.
3. Plant operators must be certified.
4. Plant owners must notify SERM when chlorine levels are low and when equipment breaks down.
5. Watershed planning will be used to protect source water from contamination.

Lessons Learned

The engineering lessons are basic:

1. The outbreak prompted the Province to change the way it regulated water and wastewater municipal utilities.
2. Don't locate a drinking water treatment inlet immediately downstream of wastewater effluent.
3. Raw source water should come from a protected watershed, not from a large turbid waterway downstream from cities and agricultural areas.
4. A multi-barrier approach is necessary to protect from Cryptosporidium.
   a) Protect the watershed from intensive agriculture.
   b) Treat the organisms like particles using flocculation and sedimentation
   c) Use a combination disinfection procedure using UV light and chlorine.

From an organizational perspective, the following are essential:

1. Management must emphasize sound technical practices.
2. Optimum allocation of resources must be made, including access to funds for small towns like North Battleford.
3. Prompt, effective communication among all government agencies is necessary.

References

Engineering Section, Department of Public Works and Utilities. (2006). *City of North Battleford 2005 Distribution Water Report, City of North Battleford*, Available at http://www.cityofnb.ca/noticefiles/2549_DRINKINGWATERReport2005.pdf , (accessed on December 14, 2006).

Fewtrell, L., MacGill, S.M., Kay, D., and Casemore, D. (2001). "Uncertainties in Risk Assessment for the Determination of Drinking Water Pollutant Concentrations: Cryptosporidium Case Study," *Water Research,* 35(2), 441-447.

Finch, G.R. and Belosevic, M. (2001). "Controlling *Giardia* spp. and *Cryptosporidium* spp. in Drinking Water by Microbial Reduction Processes," *Canadian Journal of Civil Engineering*, 28 (Suppl. 1), 67-80.

Hrudley, S.E., Payment, P., Huck, P.M., Gillham, R.W., and Hrudley, E.J. (2003). "A Fatal Waterborne Disease Epidemic in Walkerton, Ontario: Comparison with Other Waterborne Outbreaks in the Developed World," *Water Science and Technology* 47, 7-14.

Hsu, B., Huang, C., Hsu, Y., Jiang, G., and Hsu, C.L. (2001). "Evaluation of Two Concentration Methods for Detecting Giardia and *Cryptosporidium* in Water," *Water Research*, 35(2), 419-424.

Hsu, B. and Yeh, H.H. (2003). "Removal of Giardia and *Cryptosporidium* in Drinking Water Treatment: A Pilot-Scale Study," *Water Research*, 37, 1111-1117.

Jameson, P., Hung, Y., Kuo, C., and Bosela, P. (2008). "Cryptosporidium Outbreak (Water Treatment Failure): North Battleford, Saskatchewan – Spring 2001," *Journal of Performance of Constructed Facilities*, 22 (5), 342-347.

Laing, R.D. (2002). *Report of the Commission of Inquiry. The North Battleford Water Inquiry*. Regina, SK, 1-372.

Madore, M.S., Rose, J.B., Gerba, C.P., Arrowood, M.J., and Sterling, C.R. (1987). "Occurrence of Cryptosporidium Oocysts in Sewage Effluents and Selected Surface Waters," *Journal of Parasitology*, 73(4), 702-705.

Medema, G.J., Schets, F.M., Teunis, P.F.M., and Havelaar, A.H. (1998). "Sedimentation of Free and Attached *Cryptosporidium* Oocysts and *Giardia* Cysts in Water," *Applied and Environmental Microbiology*, 64(11), 4460-4466.

O'Connor, D.R. (2002). *Report of the Walkerton Commission of Inquiry. The Walkerton Inquiry*. Toronto, ON, 1-504.

O'Donoghue, P. J. (1995). "*Cryptosporidium* and Cryptosporidiosis in Man and Animals," *International Journal for Parasitology*, 25(2), 139-195.

Peeters, J.E., Mazas, E.A., Masschelein, W.J., De Maturana, I.V.M., and DeBacker, E. (1989). "Effect of Disinfection of Drinking Water with Ozone or Chlorine Dioxide on Survival of *Cryptosporidium parvum* Oocysts," *Applied and Environmental Microbiology*, 55(6), 1519-1522.

Peterson, H. (2004). "Drinking Water Treatment: Where are we Heading?" Available at <http://www.safewater.org> (accessed on September 20, 2006).

Searcy, K.E., Packman, A.I., Atwill, E.R., and Harter, T. (2005). "Association of *Cryptosporidium parvum* with Suspended Particles: Impact on Oocyst Sedimentation," *Applied and Environmental Microbiology*, 71(2), 1072-1078.

Stirling, R., Aramini, J., Ellis, A., Lim, G., Meyers, R., Fleury, M., and Werker, D. (2001). "Waterborne Cryptosporidiosis Outbreak, North Battleford, Saskatchewan, Spring 2001," *Canadian Communicable Diseases Report*, 27(22), 185-192.

Stirling, R., Aramini, J., Ellis, A., Lim, G., Meyers, R., Fleury, M., and Werker, D. (2001). "North Battleford, Saskatchewan, Spring 2001 Waterborne Cryptosporidiosis Outbreak," *Health Canada*, 1-22.

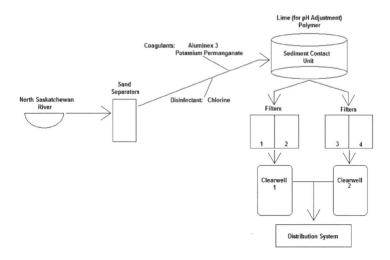

Figure 3-2. **Process Diagram of North Battleford's Surface Water Plant during Cryptosporidium Outbreak.**
Source: Jameson, et.al. (2008), ©ASCE

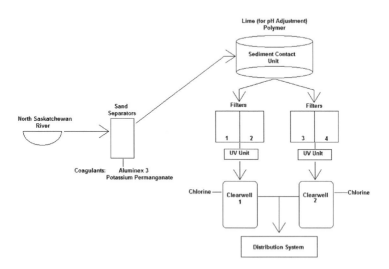

Figure 3-3. **Process Diagram of North Battleford's Surface Water Treatment Plant Post-Remediation.**
Source: Jameson, et al. (2008), ©ASCE

Chapter 4

Bridge Failures

ASHTABULA BRIDGE
(1876)

The Ashtabula Bridge was constructed during 1863-1865 over a stream in Ashtabula, Ohio. The bridge collapsed on December 29, 1876 after eleven years of service. The conceptual design of the Ashtabula Bridge was carried out by Amasa Stone, the president of the Cleveland, Painesville and Ashtabula Railroad, which later became a part of the Lake Shore and Michigan Southern Railway. The bridge was a simply supported, parallel chord, Howe truss with a span of 47 m (154 ft) consisting of 14 panels of 3.4 m (11 ft) each. The depth of the truss was 6 m (20 ft) and the center-to-center spacing of the trusses was 5 m (17 ft). The bridge was a deck structure carrying double track railway. Stone employed Joseph Tomlinson to determine the member sizes and prepare the fabrication drawings. Tomlinson carried out the task assigned to him and supervised the entire fabrication. However, he did not participate in the supervision of the construction.

The diagonals subjected to compressive forces consisted of several built-up iron I-sections. The top chord consisted of segments, two panels long, fitting between the lugs on the iron casting. The beams carrying the timber floor rested directly on the top chord, thus introducing flexure in the top chord of the truss. The position of each track was such that the live load of one train was carried predominantly by the truss near to that track.

Failure of the bridge took place at about 7:30 p.m. on December 29, 1876 during a severe snowstorm. A train with two steam locomotives was crossing the bridge heading west at an estimated speed of 20 to 25 km per hour (12 to 15 miles per hour). As the first locomotive was about to complete the crossing, the bridge began to fail. The second locomotive, its coal tender and eleven cars fell 20 m (65 ft) into Ashtabula Creek. The accident was a national tragedy causing 80 deaths.

On January 12, 1877, the Legislature of Ohio appointed a joint committee to investigate the cause of failure. A report to the legislature was made on January 30, 1877. In addition, the coroner's jury appointed an engineer to investigate the causes of the disaster. A third major investigation of the failure was carried out by Charles MacDonald.

Each investigation independently concluded that the failure occurred in the second and third panels of the south truss, though the investigators were undecided whether it was the top chord, or the compressive brace that initiated the failure. The Joint Committee concluded the bridge failed because of inadequate inspection. The coroner's jury made the same point, adding that iron bridges were in their infancy and that an experiment should not have been made on a bridge with such a deep chasm. Charles MacDonald suggested that a fatigue crack originated at a flaw in the lug and propagated under repeated stress cycles.

Lessons Learned

The failure bolstered the call for consulting bridge engineers and standard design specifications. It raised the awareness of the reliability of iron castings.

References

Gasparini, D.A. and Fields, M. (1993). "Collapse of Ashtabula Bridge on December 29, 1876," *Journal of Performance of Constructed Facilities* 7(2), 109-125.
McDonald, C. (1877). "The Failure of Ashtabula Bridge," *ASCE Transaction*, 6, 74-87.
"Report of the Joint Committee Concerning the Ashtabula Bridge Disaster," (1877). Nevins Myers State Printers, Columbus, OH.

TAY BRIDGE
(1879)

Spanning the Firth of Tay at Dundee, Scotland, the Tay Bridge was opened to railway traffic on May 31, 1878. The bridge collapsed on December 28, 1879. With a length of 3146 m (10,321 ft), when completed it was the longest bridge in the world ever built over a stream. The bridge was made of eighty-five simply-supported spans, with six at 8.2 m (27 ft), fourteen at 20.6 m (67.5 ft), fourteen at 21.5 m (70.5 ft), two at 26.8 m (88 ft), twenty-one at 39.3 m (129 ft), thirteen at 44.5 m (146 ft), one at 49.4 m (162 ft), one at 51.8 m (170 ft) and thirteen at 74.7 m (245 ft). The shorter spans were deck trusses, the longer ones were through trusses and the 51.8 m (170 ft) span section was a bowstring truss.

Many difficulties were encountered during the construction. The Tay, being a tidal river, was subjected to currents and exposed to wind gusts that constantly hampered the construction operations. Many changes were introduced as the construction progressed and the final structure differed in many ways from what was originally planned. The modifications were not recorded.

The bridge failed during a major storm at the estuary of the Tay. A train, traveling north from Edinburgh to Dundee, had to cross the bridge just before reaching Dundee, but never reached its destination. It is not known whether the bridge collapsed over a portion of its length as the train passed over, or whether the bridge had already collapsed before the train reached it. Thirteen spans of the bridge collapsed. The collapse resulted in the death of 75 people.

Lesson Learned

The principal cause of the collapse was an improper estimate of the wind force. According to the records the wind pressure was taken to be 0.5 kPa (10 psf), but in retrospect it should have been 2.4 kPa (50 psf). There were also some shortcomings in inspection, particularly at the Wormit foundry where the bridge elements were cast. The Court of Inquiry determined that Thomas Bouch, the engineer in charge, was entirely responsible for the errors in design and principally responsible for errors in construction and maintenance.

References

The Engineer (1941). "Historic Accidents and Disasters, The Tay Bridge," *The Engineer*, 172, July 11, 18-20; July 18, 34-35.

Engineering. (1880). "Tay Bridge Inquiry," *Engineering*, 20, Apr. 23, 320-323; Apr. 30, 335-339; May 7, 363-366; May 14, 387-390.

Engineering. (1880). "The Tay Bridge Accident," *Engineering*, 20, Jan. 2, 11-13; Jan. 9, 31-34; Jan. 30, 92-94; Feb. 13, 132-133; Mar. 5, 191-193.

Engineering News. (1880). The Tay Bridge," *Engineering News*, 7, Jan. 10, 10-13; Jan. 24, 36-39; Jan. 31, 44-46, 46-47; Feb. 14, 59-60; Feb. 28, 79-84; Mar. 20., 107; Apr. 24, 146; July 31, 259-260; Aug. 21, 284.

Prebble, J. (1956). *Disaster at Dundee*, Harcourt Brace & Company, New York, NY.

THE QUEBEC BRIDGE
(1907 & 1916)

It took twenty years to design and build the Quebec Bridge over the St. Lawrence River. During its construction there were two collapses. The first collapse occurred on August 29, 1907, fifteen minutes before the end of the working day, when the south anchor arm of the bridge failed. Seventy-four workers were killed. The second failure occurred on September 11, 1916, when the center span of the newly reconstructed bridge collapsed. Another eleven deaths occurred. The bridge was opened to traffic in August 1919. The bridge was a double-track railway cantilever truss structure with clear span of 1,800 ft. At the time it was the longest bridge span ever attempted.

In the first collapse there was warning prior to the failure. The compression member in the second shoreward panel from the south pier of the anchor arm showed distortion of 5.7 cm (2.25 in.) two days prior to the collapse. The second collapse occurred when the simple span weighing 50 MN (5000 tons) was being lifted into place. The cruciform casting of the southwest corner of the span failed and allowed the corner of the span to drop on the lifting girder. The blow kicked the girder backwards causing the span to be left hanging unevenly on three corners only.

Lessons Learned

The cause of the first failure was inadequate design of compression members which buckled during the construction. However, considering the enormous size of the structure, insufficient thought was given to the design and the supervision was inadequate. The failure was a tragedy that led to the development of Canadian design specifications for bridge structures.

The second failure was triggered by the fracture of the hoisting mechanism which caused the suspended span to collapse. Engineering design is necessary for the means and methods of lifting operations.

References

Barker, H. (1917). "Quebec Bridge Suspended Span Hung from Cantilever," *Engineering News Record (ENR)*, 79, Sept. 13, 581-588.

The Engineer. (1941). "Historic Accidents and Disasters, The First Quebec Bridge," *The Engineer*, 172, Oct. 24, 266-268; Oct. 31, 286-288, 406-409.

Engineering. (1907). "The Quebec Bridge Disaster," *Engineering*, 84, Sept., 328-330.

Engineering News. (1907). "The Fall of the Quebec Cantilever Bridge," *Engineering News*, 58, Sept. 5, 258-265; letters, Sept. 19; 318-319, 319-321.

Engineering Record (ER). (1907). "The Anchor Pier Towers of the Quebec Bridge," *ER*, 55, Jan.12, 34-35.

ER. (1907). "Erection Attachments for Bottom Chords and Vertical Posts of the Quebec Bridge," *ER*, 55, Jan. 19, 71-74.

ER. (1907). "Erection of the Main Vertical Posts of Quebec Bridge," *ER*, 55, Jan. 26, 92-94.

ER. (1907). "Handling Eyebars at the Quebec Bridge," *ER*, 55, Feb. 9, 153-156.

ER. (1907). "The Traveler for the Erection of the Quebec Bridge," *ER*, 55, Feb. 23, 198-210.

ER. (1907). "Erection of South Anchor Arm of the Quebec Bridge," *ER*, 55, Mar. 2, 291-292.

ER. (1907). "The Cause of Quebec Bridge Failure," *ER*, 56, Sept 14, 276; Sept. 21, 302.

ER. (1907). "The Quebec Bridge Superstructure Details I," *ER*, 56, II, July 6, 25-27; III, July 13, 36-37; IV, July 20, 65-66; V, July 27, 89-91; VI, Aug. 3, 130-131; VII, Aug. 10, 159-160; VIII, Aug. 17, 169-170; IX, Aug. 24, 210-211.

ER. (1907). "Erection of the South Cantilever Arm of the Quebec Bridge," *ER*, 56, Sept. 28, 343-345.

ER. (1908). "Report of the Royal Commission on Quebec Bridge," *ER*, 57, April.

Horton, H.E. (1907). "Lattice Bars in the Quebec Bridge," *ER*, 56, Sept. 28, 357.

Meyers, A.J. (1916). "Quebec Bridge Suspended Span," *Engineering News*, Sept. 14, 524-529.

Roddis, W.M.K. (1991). "The 1907 Quebec Bridge Collapse: A Case Study in Engineering Ethics," *Proceedings, The National Steel Construction Conference, Washington, DC*, 23-1 to 23-11.

FALLS VIEW BRIDGE
(1938)

On January 27, 1938, the Falls View arch bridge, downstream of Niagara Falls, was torn from its foundation as a result of a severe ice jam. The bridge was a tourist attraction, known as the Honeymoon Bridge. The construction of the bridge started in 1895 and was completed in 1898. The structure was a two-ribbed steel arch of 256 m (840 ft) span. Each rib was a two-hinged truss arch with a uniform depth of 7.9 m (26 ft) and a rise of 45.7 m (150 ft). The chord members were plate and angle box sections. Most of the other members were steel sections with lattice connections. The original wooden deck of 14 m (46 ft) width carried a two-track street railway and was supported by unbraced single spandrels. The four concrete and stone foundations rested in solid rock approximately 12 m (40 ft) above the normal water level.

The ice jam was formed during the night of January 25, 1938. By the following afternoon, ice had piled up to a height of 15 m (50 ft) above normal river level, or 3 m (10 ft) above the pins supporting the arch. The ice pack moved downstream like a glacier for about 122 m (400 ft) covering at least 9 m (30 ft) of the upstream truss, causing the failure of many of the bracing members. Shortly thereafter the structure was closed to traffic. The movement of the ice pack was halted, but the upstream truss continued to move very slowly downstream accompanied by further buckling and failure of secondary members. On the afternoon of January 27, the buckled section of the lower chord broke with loud report and the bridge collapsed. The bridge was replaced by the Rainbow Arch, a fixed arch rib of 290 m (950 ft) span.

Lessons Learned

The principal cause of the failure was the proximity of the ice mass to the structure and the flexibility of the structure. Although it is not always possible to design for unusual natural phenomena, the foundation of bridges should be protected where possible.

References

Buck, R.S. (1938). "Niagara Arch Memories," *Engineering News Record*, 120, February 24, 297-298.

ENR.(1938). "Cable and Deck Salvaged from Falls View Bridge," *Engineering News Record*, 120, February 24, 311.

ENR.(1938). "Falls View Bridge Sinks in River," *Engineering News Record*, 120, 559.

ENR.(1938). "Ice Power," *Engineering News Record*, 120, 171.

ENR.(1938). "Record Ice Jam at Niagara Falls Wrecks Famous Arch Bridge," *Engineering News Record*, 120, February 3, 161,168-169.

SANDO ARCH
(1939)

On August 31, 1939, during the construction of the reinforced concrete arch bridge over the Angerman River between the villages of Sando and Lunde in Sweden, the timber centering of the forms collapsed. The failure was a major disaster, costing the lives of eighteen construction workers.

The bridge with its span of 264 m (866 ft) was to have become the reinforced concrete arch with the longest span. The arch rib was designed as a three-cell hollow rectangular section having a width of 9.4 m (31 ft) and a depth of 2.7 m (8.75 ft) at the crown and 4.5 m (14.75 ft) at the abutment. The external walls of the hollow section were 30 cm (11-7/8 in.) thick and the internal walls were 20 cm (7-7/8 in.) thick.

The construction schedule called for casting the bottom slab first, followed by the walls and the top slab. The bottom slab was cast in sections and each section was allowed to set before filling the sections in between. At the time of the accident parts of the bottom slab had been completed and the workmen were cleaning the forms in preparation of pouring the center section when the timber centering failed.

The report of the investigating committee appointed after the disaster states that the failure of the timber was caused by the insufficient strength and stiffness of the transverse bracing between the two flanges of the arch. However, later investigations revealed that lateral instability of the arch was primarily responsible for the failure. The Sando Arch was rebuilt and opened to traffic August 28, 1943.

Lessons Learned

Concrete structures often require complex formwork that must act as the primary structure during construction. Proper design, construction, and removal of timber formwork must be executed with the same care as the design of the permanent structure and with full consideration of anticipated loads throughout the construction process.

References

Die Sando-Strassenbrucke in Schweden, Schweizerische Bauzeitung (1943). 119, July 16, 105.

ENR. (1939). "This Would Have Been World's Longest Concrete Arch," *Engineering News Record*, September 28, 34.

Granholm, H. (1961). "Sandobrons Bagstallining," *Transactions of Chalmers University of Technology*, Gothenburg, Sweden, No. 239.

Remarques sur la resistance des cintres en bois de grande ouverture, (1940). L'effondrement du cintre du pont en beton arme sur le fleuve Angerman, en Suede, le Genie Civil, April 27, 283.

Ros, M. (1940). "Zum Lehrgerust-Einsturz der Sando-Brucke Uber den Angermanalv in Schweden," Schweizerische sauzeitung, 115, January 20, 27-32.

TACOMA NARROWS BRIDGE
(1940)

The Tacoma Narrows suspension bridge, nicknamed "Galloping Gertie" for its propensity to oscillate, opened for traffic on July 1, 1940, and was the most flexible suspension bridge of its time. After being in service for only 129 days, the bridge failed late in the morning of November 7, 1940. Under a 68 km/h (42 mph) wind, the bridge, which normally vibrated in a vertical plane, began to oscillate with the opposite sides out of phase (torsional mode). The oscillation became extremely violent, until the failure began at mid span, with buckling of the stiffening girders and lateral bracing. The suspenders snapped, and sections of the floor system fell to the water below. Almost the entire suspended span between the towers fell into the water. The side spans, which remained, sagged about 9 m (30 ft) bending back the towers sharply by the pull of the side span cables. The towers, which were fixed at the base by steel anchors deeply embedded in concrete piers, survived but were damaged.

The dramatic collapse of this bridge, captured on film by Professor F. B. Farquharson of the University of Washington, Seattle, who was on the main span just before the failure taking motion pictures of the violent twisting of the bridge deck, provides a spectacular example of the forces developed when dynamic resonance occurs. Because of its excessive motions, the bridge was closed to traffic sometime before the failure occurred. Hence, there was no loss of human life.

The contract drawings consisted of 39 sheets and were signed by the highway and bridge engineers of the State of Washington. The drawings carried the signatures of Moran, Proctor and Freeman, the consultants on substructure and of Leon Moisseiff as consultant of superstructure. In addition, the engineer retained by the Reconstruction Finance Corporation to review the plans commented on the flexibility of the suspended structure, but deferred to the wider experience of Moisseiff. The design was also reviewed and approved by a board of consultants appointed by the Washington Toll Bridge Authority.

The total cost amounted to $6,400,000 of which $2,880,000 was a grant from the Public Works Administration and $3,520,000 was a loan from the Reconstruction Finance Corporation to be repaid by the tolls.

The suspended span was 853 m (2,800 ft) and the two side spans were 335 m (1,100 ft) each. The 853 m (2,800 ft) main span made this the third longest suspension bridge of its time. The roadway was 7.9 m (26 ft) wide and the sidewalks were 1.5 m (5 ft) wide on each side. The spacing between the cables and stiffening girders was 11.9 m (39 ft), a 1:12 span ratio. The main piers were 19.7 m (64.5 ft) by 35.8 m (117.5 ft) in plan. The towers were 128 m (420 ft) in height. Each cable consisted of nineteen strands of 332-No. 6 cold drawn galvanized wires. The diameter of each cable under wrapping was 435 mm (17 1/8 in.) with a net area of 1228 sq. cm (190 sq. in.). The original design specified stiffening trusses, but was changed to stiffening girders 2.4 m (8 ft) deep, a 1:350 span ratio.

Based upon previous experience with suspension bridges with shallow stiffening girders, the engineers anticipated oscillations. From the time it was constructed, however, the bridge developed a reputation for disturbing oscillations, up to 1.3 m (50 in) in amplitude. For the most part, these oscillations were not considered to be dangerous. Studies of the bridge vibration problem were undertaken, including wind tunnel testing of models at the University of Washington at Seattle, and modifications were made, such as the installation of cable ties attached to concrete anchors. These cables broke three or four weeks before the collapse. Deflector vanes to change the aerodynamic characteristics had also been developed, but their installation was under negotiation at the time of the collapse. After the failure, a board of engineers was appointed by the Administrator of the Federal Works Agency to determine the causes of failure.

Lessons Learned

The Board of Engineers concluded that the bridge was well designed and built to resist safely all static forces. Its failure resulted from excessive oscillations made possible by the extraordinary degree of flexibility of the structure. The Board determined with reasonable certainty that the first failure was the slipping of the cable band on the north side of the bridge to which the center ties were connected. This slipping may have initiated the torsional oscillations. The Board recommended more studies to understand the aerodynamic forces acting on suspension bridges.

Thus, incompetence or neglect was not the cause. The failure was due to the torsional oscillations made possible by the narrow width and small vertical rigidity of the structure. Those actions and forces were previously ignored or deemed to be unimportant in suspension bridge design. This failure emphasized the need to consider aerodynamic effects in the design of a suspension bridge. Modern bridge decks are designed to eliminate the aero-elastic instability that pushed Galloping Gertie to failure.

References

Ammann, Othmar H., von Karman, Theodore and Woodruff, Glenn B. (1941). *The Failure of the Tacoma Narrows Bridge, A Report to the Administrator*, Federal Works Agency, Washington D.C., March 28.

Andrew, Charles E. (1943). "Observations of a Bridge Cable Unspinner," *ENR*, August 26, 89-91.

Andrew, Charles E. (1943), "Unspinning Tacoma Narrows Bridge Cables," *ENR*, January 14, 103-105.

Andrew, Charles E. (1945). "Redesign of Tacoma Narrows Bridge," *ENR*, Nov. 29, 64-67, 69.

Bowers, N. A. (1940). "Tacoma Narrows Bridge Wrecked by Wind," *ENR*, November 14, 1, 10, 11-12.

Bowers, N. A. (1940). "Model Tests Showed Aerodynamic Instability of Tacoma Narrows Bridge," *ENR*, November 21, 40-47.

Delatte, Norbert J. (2009). *Beyond Failure: Forensic Case Studies for Civil Engineers*, ASCE Press, 26-37.

ENR. (1940). "Tacoma Narrows Bridge Wrecked by Wind," *ENR*, Nov. 14, 647.

ENR. (1940). "Model Tests Showed Aerodynamic Instability of Tacoma Narrows Bridge," *ENR*, 124, 674.

ENR. (1940). "Tacoma Narrows Bridge Being Studied by Model," *ENR*, 125, 139.
ENR. (1940). "Aerodynamic Stability of Suspension Bridges," *ENR*, 125, 670.
ENR. (1940). "Dynamic Wind Destruction," *ENR*, 125, 672-673.
ENR. (1940). "Board Named to Study Tacoma Bridge Collapse," *ENR*, 125, 733.
ENR. (1940). "Another Consultant Board Named for Tacoma Span," *ENR*, 125, 735.
ENR. (1940). "Tacoma Narrows Collapse," *ENR*, 125, 808.
ENR. (1940). "Action of Von Karman Vortices," *ENR*, 125, 808.
ENR. (1940). "Resonance Effects of Wind," *ENR*, 125, 808.
ENR. (1940). "Comment and Discussion," *ENR*, Nov. 21, 40.
ENR. (1940). "Comments and Discussions," *ENR*, Dec. 5, 40-41.
ENR. (1941). "Why the Tacoma Narrows Bridge Failed," *ENR*, 126, 743-747.
ENR. (1943). "Tons of Scrap Wire from Tacoma Narrows Bridge," *ENR*, Mar. 11, 87.
ENR. (1943). "Tons of Scrap Wire from Tacoma Narrows Bridge," *ENR*, August 26, 89-91.
Farquharson, F. B. (1940). "Dynamic Models of Tacoma Narrows Bridge," Civil Engineering, Volume 10, September, 445-447.
Farquharson, F.B. (1946). "Lessons in Bridge Design Taught by Aerodynamics Studies," Civil Engineering, Vol. 16, Aug., 344-345.
Farquharson, F.B. (1950). "Aerodynamic Stability of Suspension Bridges with Special Reference to the Tacoma Narrows Bridge, Investigations Prior to 1941," University of Washington Engineering Experiment Station Bulletin 116, Part I.
Farquharson, F.B. (1952). "Aerodynamic Stability of Suspension Bridges with Special Reference to the Tacoma Narrows Bridge, the Investigation of the Models of the Original Tacoma Narrows Bridge under the Action of Wind," University of Washington Engineering Experiment Station Bulletin 116, Part III, June.
Farquharson, F.B. (1954). "Aerodynamic Stability of Suspension Bridges with Special Reference to the Tacoma Narrow Bridge, Model Investigations which Influenced the Design of the Tacoma Narrows Bridge," University of Washington Engineering Experiment Station Bulletin 116, Part IV, April.
Goller, R.R. (1965). "The Legacy of Galloping Gertie 25 Years Later," CE, 35, Oct., 50-53.
Scott, R. (2001). "In the wake of Tacoma: suspension bridges and the quest for aerodynamic stability," ASCE Press.
Smith, Frederick C., Vincent, George S. (1950). "Aerodynamic Stability of Suspension Bridges with Special Reference to the Tacoma Narrows Bridge, Mathematical Analysis," University of Washington Engineering Experiment Station Bulletin 116, Part II, October.
Vincent, George S., (1951). "Suspension Bridge Vibration Formulas," *ENR*, Jan. 11, 32-34.
Vincent, George S., (1954). "Aerodynamic Stability of Suspension Bridges with Special Reference to the Tacoma Narrows Bridge, Extended Studies: Logarithmic Decrement Field Damping, Prototype Predictions, Four Other Bridges," University of Washington Engineering Experiment Station Bulletin 116, Part V.

PEACE RIVER BRIDGE
(1957)

The Peace River Bridge on the Alcan Highway in British Columbia failed on October 16, 1957, when the north concrete anchorage block moved forward some 3.7 m (12 ft) on its shale base.

The Peace River Bridge was part of a rush wartime program to complete the Alcan Highway connecting the United States with Alaska. The suspension bridge had a main span of 283 m (930 ft) and the side spans between the towers and cable bents were 142 m (465 ft) each. Simple truss spans connected the cable bents to the anchorages. The roadway was 7.3 m (24 ft) wide and the center-to-center spacing of the cables was 9 m (30 ft). The cables, made of twenty four 5 cm (1 7/8 in.) strands, were arranged in rectangular form with dimensions of 15 cm by 10 cm (6 in. by 4 in.). The stiffening trusses were 4 m (13 ft) deep.

The Bridge was designed and constructed by the Bureau of Public Roads, the predecessor of the Public Roads Administration. Because of the rush nature of the job, no piling was used to support the anchorages. The sliding of the anchorage on the shale base caused slacking off of the main cables, tipped over the cable bent, dropped the side span suddenly, ripping loose from its 6 cm (2.5 in.) hangers. The first indication that the anchorage was moving came about 12 hours before the collapse when the water supply line crossing the bridge for the new scrubbing plant of the Pacific Petroleum Company was cut. The bridge was immediately closed to traffic. A large crowd gathered to witness the collapse, which was thoroughly photographed. The Canadian Army Engineers put a small ferry 16 km (10 miles) downstream to provide essential transportation for Yukon and Alaska.

Lessons Learned

In order for a suspension bridge to support the applied loads in the intended manner, it is essential that the anchorages be securely fixed to the ground. Any horizontal motion of an anchorage will cause slackening of the cables with the possibility of collapse of the structure. The Peace River anchorages were supported on footings which did not stay fixed at the intended location.

References

Birdstall, Blair (1944). "Construction Record Set on Alcan Suspension Bridge," *ENR*, January 13, 26.

ENR. (1957). "Anchorage Slip Wrecks Suspension Bridge," (1957). *ENR*, October 24, 26.

ENR. (1958). "New Bridge for Peace River on Alcan Highway Planned," *ENR*, January 23, 17.

ENR. (1958). "Peace River Bridge Failure Was Avoidable," *ENR*, February 13, 10, 16.

ENR. (1959). "Canada Works on Its Two Fallen Bridges," *ENR*, March 19, 49.

THE SECOND NARROWS BRIDGE
(1958)

On June 17, 1958, the falsework of the partly completed Second Narrows Bridge in Vancouver, British Columbia buckled and plunged two spans of the bridge into Burrard Bay. Fifteen men died in the collapse and twenty were injured. The six-lane cantilever truss bridge was to be an important link between the cities of Vancouver and North Vancouver. The main cantilever structure was 620 m (2034 ft) long consisting of a 335 m (1,100 ft) cantilever span, and two 142 m (467 ft) anchor spans. In addition the bridge had four 87 m (285 ft) steel truss and nine 37 m (120 ft) prestressed concrete approach spans. The construction of the bridge began in February 1956 and was to be completed by the end of 1958 at a cost of $16 million.

The two sections that fell were the partly erected north anchor span and a completed simple truss span adjacent to it. The workers were moving additional steel to the overhanging end when the collapse occurred. The bent supporting most of the 20 MN (2,000 ton) anchor span buckled, dropping one end of the span into the water. The impact moved the top of the permanent concrete pier by a few ft, plunging the adjacent simple span into the water.

Lessons Learned

To determine the reason for the failure of the temporary supports, an investigation was carried out under the British Columbia's Supreme Court Chief Justice, Sherwood Leu. The investigation revealed that the bent supporting the cantilever bridge section was not properly designed. The grillage was designed by comparatively inexperienced engineers without effectively checking the calculations. The bridge was completed in July 1960.

References

ENR. (1958). "What Happened at Vancouver? Probers Seek Cause of Bridge Collapse," *ENR*, June 26, 21-22.
ENR. (1958). "What Happened in Vancouver?," *ENR*, July 3, 100.
ENR. (1958). "Bridge Crash Witnesses Testify," *ENR*, July 31, 24.
ENR. (1958). "Faulty Grillage Felled Narrows Bridge," *ENR*, October 9, 24.
ENR. (1958). "Well Done, British Columbia," *ENR*, Oct. 23, 100.
ENR. (1958). "Report Blames Contractor for Vancouver Failure," *ENR*, December 11, 32.
ENR. (1958). "Vancouver Bridge Failure Probed Properly," *ENR*, December 18, 100.
ENR. (1960). "Vancouver Spans Near Completion," *ENR*, Jun 30, 24.

KING STREET BRIDGE
(1962)

The King Street Bridge was an all-welded steel girder structure consisting of three main sections, a high level section and two lower level spans which flanked both sides of the high level portion. The spans served to carry roadways over the Yarra River and were completed on April 12, 1961. On the morning of July 10, 1962, brittle fracture failure occurred at points 4.9 m (16 ft) from the ends of one of the 30 m (100 ft) long approach spans under a load of 470 kN (47 tons), which was within the permissible design limits for the bridge, at a temperature of -1 degree Centigrade (30 degrees Fahrenheit). Three of the four girders fractured at points 4.9 m (16 ft) from both the southern and northern ends whereas the fourth one failed only at one position, namely 4.9 m (16 ft) from the southern end. The failure of the four girders was attributed to a combination of three factors: inappropriate steel for welding, unsatisfactory design details and low ambient temperatures.

The steel used, British Standard 968.1961, is similar to ASTM A 440 and was commonly used in riveted and bolted construction. Welding of such high carbon steel often results in weaknesses being generated in the heat affected zones and the triggering of lamellar tearing failures. Lack of preheating in the short transverse welds at the ends of the cover plates which terminated at the position of fracture is thought to have contributed to crack initiation.

The thickening of the flanges at the points of maximum tensile stress by the addition of cover plates was not a favorable design feature. The temperature on the day of collapse was below that at which the transition from ductile to brittle steel characteristics occurs. Brittle behavior favors crack initiation and propagation by increasing the stress intensity factor at any surface or interior flaws. These conditions were found to contribute to the failure of the King Street Bridge. The bridge was repaired by externally prestressing the girders with steel cables.

Lessons Learned

Inappropriate steel selection, undesirable design details and unusually low temperatures were the main contributory factors leading to the failure of the King Street Bridge. While the low temperature could not have been avoided, the other aspects were within the control of those involved in the bridge's inception.

References

Engineering. (1962). "Brittle Fracture of an Alloy Steel Bridge," *Engineering*, 194, 5031, Sept 21, 375.
ENR. (1962). "Steel Blamed in Bridge Failure," *ENR*, 169, 12, Sept. 12, 139.

POINT PLEASANT BRIDGE—SILVER BRIDGE
(1967)

The Point Pleasant I-bar suspension bridge between Point Pleasant, West Virginia and Kanagua, Ohio which was built in 1928, failed at 5:00 p. m. on December 15, 1967. Forty-six people died in the accident and thirty-seven vehicles on the bridge fell with the bridge.

The center span was 213 m (700 ft) long and the side spans were 116 m (380 ft) each. The bridge was unique in that the stiffening trusses of both the center span and the two side spans were framed into the eyebar chain to make up part of the stiffening truss.

Investigation of the failure indicated that the collapse of the Point Pleasant Bridge was caused by a defective eyebar at joint 13 of the north chain, approximately 15 m (50 ft) west of the Ohio Tower. The bar which connected Joint 11 to Joint 13 developed a cleavage fracture in the lower portion of its head. Once the continuity of the suspension system was destroyed the bridge collapsed suddenly.

Lessons Learned

An eyebar suspension bridge is not a redundant structure. Failure of one eyebar is sufficient to cause collapse. If a bridge of this type is to be constructed, close and frequent inspection of the structure is necessary.

The tragedy of the failure of the Pleasant Point Bridge led to the national policy for bridge inspections. In 1968 the United States Congress enacted the National Bridge Inspection Standards (NBIS).

References

Ballard, W.T. (1929). "An Eyebar Suspension Span for the Ohio River," *ENR*, June 20, 997-1001.

Bennett, J. A. and Mindlin, Harold (1973). "Metallurgical Aspects of the Failure of Point Pleasant Bridge," *Journal of Testing Evaluation*, ASTM, I(.2), 152-161.

Dicker, Daniel (1971). "Point Pleasant Bridge Collapse Mechanism Analyzed," *CE*, July, 61-66.

ENR. (1967). "Collapse May Never be Solved," *ENR*, December 21, 69-71.

ENR. (1968). "Possible Key to Failure Found," *ENR*, January 1, 27-28.

ENR. (1968). "Point Pleasant Bridge Failure Triggers Rash of Studies," *ENR*, January 4, 18.

ENR. (1968). "Collapsed Silver Bridge is Reassembled," *ENR*, April 25, 28-30.

ENR. (1969). "Bridge Failure Probe Shuts Twin," *ENR*, January 9, 17.

National Transportation Safety Board (NTSB). (1968). *Collapse of US 35 Highway Bridge, Point Pleasant, West Virginia, December 15,1967 Highway Accident Report*, NTSB, Washington, D. C., October 4.

NTSB. (1970). *Collapse of US 35 Highway Bridge, Point Pleasant, West Virginia, December 15, 1967 Highway Accident Report*, NTSB, Washington, D. C. December 16.

Scheffey, Charles F. (1971). "Point Pleasant Bridge Collapse, Conclusions of the Federal Study," CE, July, 41-45.

Shermer, Carl (1968). "Eye-Bar Bridges and the Silver Bridge Disaster," *Engineers Joint Council*, Vol. IX, No.1, January-February, 20-31.

Steinman, D.B. (1924). "Design of Florianopolis Suspension Bridge," *ENR*, November 13, 780-782.

Time Magazine. (1967). "Disaster," *Time Magazine*, December 22, 20.

ANTELOPE VALLEY FREEWAY
INTERCHANGE
(1971 & 1994)

A few structural failures can be considered milestones in that they have had a far reaching impact on design codes and construction techniques. One such failure was the collapse of the Interstate 5/14 Freeway south connector overcrossing during the 1971 Sylmar earthquake. The overpass was in the final stage of construction, of prestressed concrete box girder design, 411 m (1349 ft) long over nine spans, having a section 10.3 m (34 ft) wide and 2.1 m (7 ft) deep. The longest column of the overpass, which was 42.7 m (140 ft) high, had an octagonal section 1.8 m by 3 m (6 ft by 10 ft) with no enlargement where the column intersected with the bottom of the beam section. The foundation for this column consisted of a 6 m (20 ft) deep, 2.4 m (8 ft) diameter cast in place drilled concrete shaft founded onto bedrock. Reinforcement for the column consisted of fifty-two 57 mm (No. 18) bars longitudinally, tied by 13 mm (No. 4 bars) at 30 cm (12 in.) on center. This column supported the center of a 117 m (384 ft) long section of the overpass which was connected to the rest of the bridge by way of two shear key type hinges on both ends of the box girder section. The shear keys were 17.8 cm (7 in.) deep vertically and 35 cm (14 in.) long. The sections were also tied together by three 3.8 cm (1.5 in.) diameter steel bolts that were added to equalize the longitudinal deflections in the superstructure arising from creep and temperature effects. This section was different from the rest of the bridge in that it was supported by one column instead of two.

On February 9, 1971, at 6:01 a.m. an earthquake assessed at Richter magnitude 6.6 occurred in the mountains behind Sylmar. The interchange suffered horizontal accelerations that were estimated as high as 0.6g. The 10 to 15 seconds of strong motion caused the superstructure of the 117 m (384 ft) section of the overpass to jump out of the shear key seats and induced the column and bridge deck to act as an inverted pendulum. The capacity of the column was found inadequate and it failed in bending at the base.

It was generally agreed that the overpass was of superior construction and did not fail as a result of any defects in workmanship or construction techniques. Prestressing elements survived the earthquake loading well and were intact in the debris.

As a result of this experience, significant changes in bridge design criteria were made including very large increases in beam seat sizes to allow for much greater longitudinal and lateral horizontal movements, the requirement for placement of hinges so that there are at least two columns between adjacent hinges along the bridge, the incorporation of spiral reinforcement to confine the longitudinal steel within the columns, the elimination of lap slices at the base of the columns, the reduction of skews in overpass structures, the increase in the amount of reinforcement at the column / deck connection to provide greater resistance to punching shear, and the elimination of the use of rocker type bearings.

On January 17, 1994, the Richter Magnitude 6.4 Northridge earthquake again caused failure of portions of this interchange (Figure 4-1). On this second occasion, some of the most severe damage occurred to sections that had been repaired following the 1971 earthquake and in other instances spans that had been under construction in 1971 failed this time. The fact that some spans were supported on columns of greatly dissimilar heights was thought to have contributed to the failures. The shorter columns, being much stiffer, were considered to have attracted disproportionately large shear forces resulting in their being overloaded with an inevitable subsequent domino effect. Also inadequate seat lengths at the ends of several spans contributed to collapse. Apparently the interchange had been scheduled for a seismic upgrade but the 1994 earthquake occurred before this had been started.

Lessons Learned

The failure of the Interstate 5/14 interchange in 1971 represented a turning point in seismic design of freeway bridges and prompted a radical change in the seismic design provisions for such structures. However these changes were not applied to the I-5/14 interchange itself. The failure in 1994 reemphasized the dangers of procrastination in undertaking seismic retrofitting once the need for such action has been established.

References

CE. (1994). "Northridge Earthquake," *CE*, 64(3), 40-47.
EERI. (1972). *Engineering Features of the San Fernando Earthquake, February 9, 1971*, Earthquake Engineering Research Laboratory Report 71-02, June, California Institute of Technology.
State University of New York at Buffalo. (1994). *The Northridge, California Earthquake of January 17, 1994: Performance of Highway Bridges*, Technical Report NCEER-94-0008, *State University of New York at Buffalo*, March.

Figure 4-1. Antelope Valley – Interstate 5 Freeway Interchange Failure – 1994.
Source: J. Dewey, USGS.

MIANUS RIVER BRIDGE
(1983)

The Mianus River Bridge forms part of the Connecticut turnpike which is 206 km (129 miles) long and crosses the state from Rhode Island to New York as part of Interstate 95. It was opened in 1958. The bridge is a six lane structure crossing the river at a 53 degree angle and is comprised of cantilever sections, suspension spans and pin-hanger connections. The deck is 19 cm (7.5 in.) thick concrete with a 5 cm (2 in.) thick asphalt top carried on steel plate girder construction, the longest span of which is 809 m (2,656 ft). The cantilever sections are projected from adjacent skewed concrete piers. The three lane suspension spans, 10.6 m by 30.5 m (35 ft by 100 ft), are held in place by two cantilever sections. One such section weighs five hundred tons and includes two nine foot deep plate girders and four stringers connected by a network of cross beams. Each span is connected at the corners by two pins and a 2 m (6.5 ft) long hanger. The pins are 17.8 cm (7 in.) long, 17.8 cm (7 in.) in diameter and each had a 2.5 cm (1 in.) diameter hole along its axis to receive an 28.5 cm (11.25 in.) long steel bolt with washers and a nut welded on each end. A pair of such pins attaches two hangers to the cantilever girder and the suspension girder. Hence, one span has a total of eight pins and eight hangers.

On June 28, 1983, in the early hours in the morning, a 30 m (100 ft) long section of the Mianus River Bridge fell taking down with it two tractor trailers and two passenger vehicles. The fallen section was a three lane east bound suspended span near the Long Island Sound entrance. The adjacent west bound section remained in place. In the investigation which followed the collapse several of the steel pins were recovered. A portion of the eighth was also found. Metallurgical tests were conducted on the fragmented pin.

An inquiry concluded that a lack of maintenance and inadequate inspection were the main causes of the collapse. It is believed that corrosion caused a pin to be pushed off the hanger which triggered the collapse.

As a result of the Mianus River bridge incident other bridges with similar design were inspected thoroughly. It was found that four other bridges on the turnpike were suffering similar deterioration thereby confirming the failure to maintain the hanging mechanism on these bridges had developed into a serious problem in other cases as well.

Lessons Learned

It was clear from this event that inadequate maintenance of major bridges can lead to catastrophic collapse. The failure of the Mianus River bridge was an example of the neglect of the infrastructure which has become a major concern to civil engineers in recent years and emphasizes the need for not only ensuring adequate standards of design and construction but also of maintenance and of service throughout the life of all structures. Many repairs and retrofits were initiated to provide increased redundancy in other bridges.

The Mianus River bridge failure investigations revealed a vulnerable characteristic of the bridge type. Adequate maintenance and repair throughout the life of all structures is necessary. the enhanced inspections that were implemented based on the information about this bridge likely prevented additional failures.

References

CE. (1985). "New Mianus River Bridge Report Disputes Earlier Study," *CE*, April, 10.

CE. (1986). "Designers Cleared in Bridge Collapse," *CE*, April, 10.

ENR. (1983). "Failed Pin Assembly Dropped Span," *ENR*, July 7, 10-12.

ENR. (1983). "Temporary Span Rushed in.," *ENR*, July 7, 11-12.

ENR. (1983). "Girder Haws Eyed in Span Collapse," *ENR*, July 14, 12.

ENR. (1983). "Hearings on Collapsed Span Focus on Rust, Skewed Plan," *ENR*, Sept. 29, 35-36.

New York Times. (1958). "Mianus River Bridge to Open," *New York Times*, July 18, 23.

New York Times. (1984). "Engineers Say Faulty Design was Factor in Mianus Bridge Collapse," *New York Times*, July 18, 23.

SAN FRANCISCO-OAKLAND BAY BRIDGE
(1989)

The two deck San Francisco/Oakland Bay Bridge was opened in 1936 and comprises two distinct structures. The West Bay Crossing from San Francisco to Yerba Buena Island is a twin suspension structure whereas the East Bay Yerba Buena to Oakland portion is composed of a series of simple span trusses and a long cantilever truss. The seismic design was based on a coefficient of 0.1 g in accordance with the standards at the time the design was completed.

On October 17, 1989 the 7.1 Richter magnitude Loma Prieta earthquake was centered about 100 km (60 miles) south of the San Francisco/Oakland bridge but caused a 15 m (50 ft) span of the upper deck to collapse onto the lower one, resulting in the death of one motorist who had the misfortune to drive into the gap.

The 4 km (2.4 mile) East Bay Crossing uses 10 bridge piers as anchor points to transfer longitudinal forces from the bridge deck to the foundation. These bridge piers vary in construction. The pier where the span collapsed is a four column, diagonally braced steel tower. It services a tributary distance of 775 m (2,544 ft) to the west and 193 m (632 ft) to the east. Since the tributary distance to west of the pier is much the larger, the longitudinal forces from the west will normally be of a greater magnitude than those from the east. To connect the two truss systems together, an expansion joint assembly consisting of I beam stringers resting on 15 cm (6 in.) wide stiffened seat supports at the west end and bolted stringer flanges to stiffened seat connections at the east end was used. The stiffened seats provided 13 cm (5 in.) of bearing support for the stringers in the pre-collapsed configuration. The upper deck floor system used four stringers which supported transverse joists carrying the concrete deck. The lower deck floor system used eleven stringers which directly supported the concrete deck. It has been estimated that during the earthquake the maximum longitudinal acceleration suffered by the eastern portion was 0.22g. The longitudinal inertia force arising from this acceleration caused the twenty-four 2.5 cm (1in.) diameter bolts used to anchor the fixed shoes to the tower columns to shear off. The truss was then free to move with the motions of the earthquake and as the deck moved eastward the stringers were pulled away from the west truss, exceeding the bearing distance of the stiffened seat and causing the stringers to fall at the west end (Figure 4-2 and 4-3). The actual easterly displacement of the shoes was measured to be at least 17.8 cm (7 in.), judging from the marks on the base plates to which the shoes had previously been attached.

The bridge was restored to use within a month of the earthquake but controversy raged for many years with respect to the optimum method of retrofit to provide greater seismic resistance than that possessed by the original structure.

Lessons Learned

The loss of a major lifeline transportation link in the San Francisco Bay area demonstrated the vulnerability of an integrated economic area to the loss of a major artery. It prompted the application of state of the art methods of earthquake engineering analysis to evaluate what seismic forces and movements long bridges are likely to experience in major earthquakes and focused attention on the need to retrofit many of the major bridges in seismic prone areas.

References

Astaneh, A. (1989). *Preliminary Report on the Seismological and Engineering Aspects of the October 17, 1989 Earthquake*, Report No. UCB/EERC-89114, Earthquake Engineering Research Center, University of California at Berkeley, Oct., 34-37.

EERI. (1990). "Loma Prieta Reconnaissance Report," Supplement to *Earthquake Spectra*, 6, Earthquake Engineering Research Institute, 162-169.

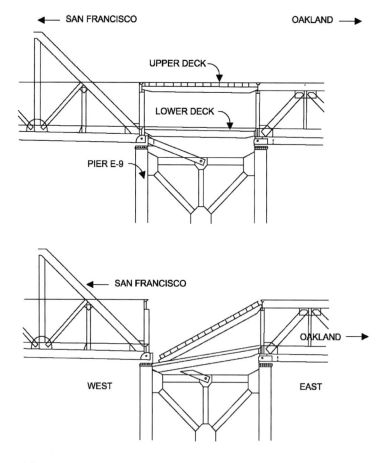

Figure 4-2. Detail of Collapse of Span of San Francisco-Oakland Bay Bridge.

Figure 4-3. Damage to San Francisco-Oakland Bay Bridge.
Source: E.V. Leyendecker, USGS.

CYPRESS VIADUCT
(1989)

The Cypress Viaduct was California's first continuous double-deck freeway structure. Construction commenced in 1955 and it was opened to traffic in June, 1957. It was situated just west of downtown Oakland and extended approximately 2.4 km (1.5 miles) in a north-south direction. Both the upper, southbound, and lower, northbound, traffic lanes were supported above ground level by a series of reinforced concrete bents. The upper frame incorporated shear keyed, essentially pinned, joints positioned so that the upper part of each bent was statically determinate.

Some recognition of the potential hazard posed by a structure which had been designed to much lower seismic load demands than would be appropriate for a structure conceived more recently had resulted in some retrofitting of the viaduct. This work comprised tying the spans together longitudinally. No strengthening of the joint areas around the shear keys was done.

The majority of the fatalities which occurred as a result of the 1989 Loma Prieta earthquake resulted from the collapse of the upper deck of the Cypress Viaduct onto the lower elevated deck, trapping and crushing vehicles in the northbound lanes. The fact that other nearby bridges and buildings survived the seismic shaking focused attention on the configuration of the Cypress Viaduct, the details of the structure and the site on which it was founded. An extensive investigation was mounted by many groups of researchers including those from the University of California, Berkeley. As a result of these investigations the characteristics of the structure and its behavior in the October 17, 1989 earthquake were clarified.

The majority of the collapsed bents followed a common pattern of failure involving slipping along a plane of weakness which existed in the stub region of the lower column to beam joint, just below the shear key at the bottom of the upper bent column (Figure 4-4). This plane of weakness in the joint region was created by closely spaced lower girder negative moment steel reinforcement which was bent down into the column. Insufficient transverse reinforcement was provided to prevent the wedge of concrete outside the plane of the bent down girder reinforcement from sliding on the sloping failure surface under the combined effects of the lateral seismic loads and the weight of the upper deck.

Lessons Learned

The reinforcement in the columns and girders was poorly detailed, even allowing for the lack of understanding, at the time the viaduct was designed, of the inelastic response demands on reinforced concrete structures in seismic zones. Confirmation of the necessity to design for realistic earthquake generated forces and displacements was one of the lessons learned from the experience of the Cypress viaduct collapse. Another was the penalty of delaying seismic strengthening once it is established that a structure is not up to current design standards.

References

EERC. (1989). *Collapse of the Cypress Street Viaduct as a Result of the Loma Prieta Earthquake*, Report No. UCBIEERC-89116, Earthquake Engineering Research Center, University of California at Berkeley.

EERI. (1990). "Loma Prieta Reconnaissance Report," Supplement to *Earthquake Spectra*, 6, Earthquake Engineering Research Institute, 151-159.

Housner, G.W. (1989). *Competing Against Time*, Report to Governor George Deukmejian on 1989 Loma Prieta Earthquake, 264 .

Figure 4-4. Cypress Viaduct Support Column Failure.
Source: H.G. Wilshire, USGS.

Chapter 5

Building Failures

AMC WAREHOUSE
(1955)

The AMC Warehouse at Wilkins Air Force Depot, Shelby, Ohio consists of 122 m (400 ft) long six-span rigid concrete frames spaced at 10.6 m (35 ft) on-center along the 61 m (200 ft) length of the building (Figure 5-1). The 10 cm (4 in.) cast-in-place gypsum roof slab is supported by prestressed concrete block purlins spanning between the rigid frames. An expansion joint runs the length of the building near the center of the 122 m (400 ft) wide frames.

Due to distress noted during construction of similar warehouses at other Air Force facilities, the Wilkins warehouse design was modified during construction in March 1954. The revision included the addition of continuous full length top bars and nominal shear stirrups throughout the span length. Approximately 18 months after casting, diagonal cracks were first noticed in the concrete frames. In August 1955 three adjacent concrete frames collapsed under dead load of the frames and roof only. All frames failed about 45 cm (18 in.) beyond the cut-off location of the negative reinforcement in the first interior span. The collapsed area was approximately 370 sq. m (4000 sq. ft).

Lessons Learned

The collapse was attributed to insufficient shear reinforcement and insufficient extension of both positive and negative reinforcing steel. The reinforcement was not able to resist the combined flexural, shear and longitudinal stresses. The longitudinal stresses (shrinkage and temperature) were believed to be higher than expected due to ineffective expansion joints. A series of load tests were performed on one-third scale model specimens of the concrete frames at the Portland Cement Association. The laboratory study was able to replicate the type of failure observed in the structure to verify the cause of the collapse. All the frames were reinforced with vertical steel straps placed around the girders of the frames to increase their shear capacity. The repair scheme was also tested at the Portland Cement Association laboratory.

Since the failure investigation found that the design of the structure conformed to current codes, the ACI Building Code was revised to limit the allowable shear stress in members without shear reinforcement to the lesser of $0.03 f_c$ or 620 kPa (90 psi). In addition, the code revision required that flexural reinforcement for continuous or restrained beams (except T-beams with an integral slab) be extended from the support to a point either 1/16 of the clear span or the depth of the member, whichever is greater, past the extreme point of inflection. The shear reinforcement was also required to carry at least two-thirds of the total shear in this region and at least two-thirds of the total shear at any section which contained negative moment reinforcement.

References

Anderson, B.G. (1957). "Rigid Frame Failures," *J. of the American Concrete Institute*, 28(7), 625-636.

Delatte, Norbert J. (2009). *Beyond Failure: Forensic Case Studies for Civil Engineers,* ASCE Press, 130-133.

Elstner, R.C. and Hognestad, E. (1957). "Laboratory Investigation of Rigid Frame Failure," *J. of the American Concrete Institute*, 28(7), 637-668.

Feld, J. (1964). *Lessons from Failures of Concrete Structures,* American Concrete Institute (ACI) Monograph No. 1, Detroit, Michigan.

Lunoe, R.R. and Willis, G.A. (1957). "Application of Steel Strap Reinforcement to Girders of Rigid Frames, Special AMC Warehouses," *J. of the American Concrete Institute*, 28(7), 669-678.

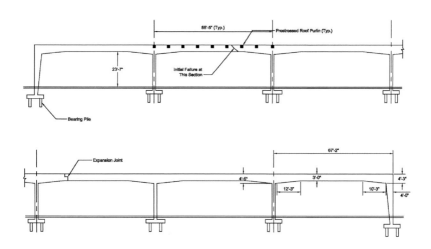

Figure 5-1. Detail of Typical Frame at AMC Warehouse.

RONAN POINT TOWER
(1968)

The need to provide replacement housing for homes destroyed in World War II prompted European development of innovative prefabricated construction techniques. One such design involved the erection of multi-story apartment buildings using factory made concrete components. The structural system comprised load bearing walls, with each level of apartments stacked directly on the one below (Figure 5-2). Floor on wall and wall on floor joints were grouted bearing surfaces (Figures 5-3 and 5-4). This was termed "system building." A high-rise apartment building at Ronan Point, Canning Town, U.K. was constructed using one such system.

On May 16, 1968 an undetected gas leak resulted in an explosion in the kitchen of a unit on the eighteenth floor when the occupant attempted to light the stove. The corner walls of this unit blew out, causing the wall above to collapse. This, in turn, impacted on the floors below and destroyed the whole corner of the building. Fourteen people were injured, three fatally.

Analysis of the event revealed that there were no alternative load paths when one part of an external wall at one level was removed. Demolition of the building also confirmed that deficiencies existed in the quality of the grouted joints between the prefabricated components.

Due to the failure of Ronan Point Tower authorities questioned the safety of other apartment towers using similar structural systems. Many were demolished well in advance of their expected life expectancy. Progressive collapse, in which local failure is followed by a chain reaction producing widespread collapse, certainly was not unknown prior to the Ronan Point event. Structures are particularly susceptible to this domino effect during of the construction process. What was unusual in the case of Ronan Point was that a relatively minor gas explosion triggered the collapse of a significant portion of a completed building.

Lessons Learned

The experience of Ronan Point reemphasized the need to be aware of the possibility of progressive collapse of constructed facilities, the desirability of providing redundancy – or fail safe possibilities – in structural systems, and the necessity of ensuring quality control in the construction process. Also, the system building technology had never been intended for buildings of this height, and had been pushed well past the limits of safety. It was disturbing that the building conformed to the codes in effect at that time in the U.K., which led to revision of those codes.

References

Allen, D.E. and Schriever, W.R. (1972). "Progressive Collapse, Abnormal Loads and Building Codes," *Structural Failure: Modes, Causes, Responsibilities,* ASCE, New York, NY.

Delatte, Norbert J. (2009). *Beyond Failure: Forensic Case Studies for Civil Engineers*, ASCE Press, Reston, VA, 97-106.

Griffiths, H., Pugsley, A. G., and Saunders, O. (1968). *Report of the inquiry into the collapse of flats at Ronan Point, Canning Town*, Her Majesty's Stationery Office, London.

Kaminetzky, D. (1991). *Design and Construction Failure*, McGraw Hill.

Levy, M. and Salvadori, M. (1992). *Why Buildings Fall Down,* Norton and Co., New York, NY, Chapter 5.

Pearson, C. and Delatte, N. (2005). "The Ronan Point Apartment Tower Collapse and its Effect on Building Codes," *Journal of Performance of Constructed Facilities*, 19(2), 172 – 177.

Petroski, H. (1994). *Design Paradigms,* Cambridge University Press.

Wearne, Phillip (2000). *Collapse: When Buildings Fall Down,* TV Books, L.L.C., New York, NY, Chapter 7.

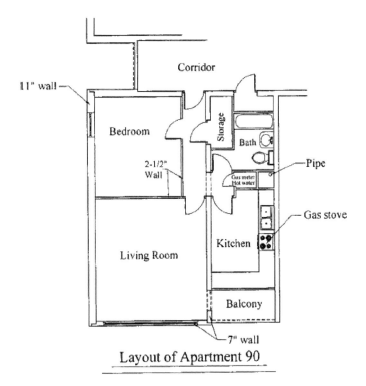

Figure 5-2. Floor Plan of Ronan Point Apartment.
Source: Pearson and Delatte (2005), ©ASCE

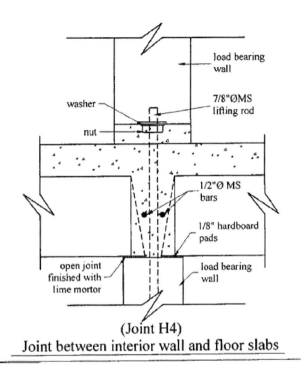

(Joint H4)
Joint between interior wall and floor slabs

Figure 5-3. Ronan Point connection.
Source: Pearson and Delatte (2005), ©ASCE

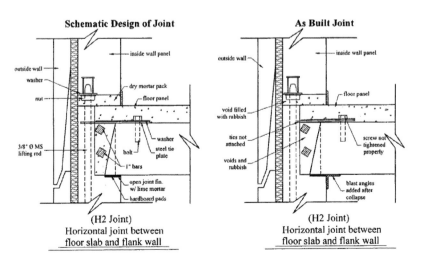

(H2 Joint)
Horizontal joint between
floor slab and flank wall

(H2 Joint)
Horizontal joint between
floor slab and flank wall

Figure 5-4. Ronan Point connection.
Source: Pearson and Delatte (2005), ©ASCE

2000 COMMONWEALTH AVENUE
(1971)

Four workers died when approximately two-thirds of a 16-story apartment building under construction in Boston collapsed on January 25, 1971. Almost 7,260 tonnes (8,000 tons) of debris were removed before the bodies of the workers could be recovered. The building had been in development for more than six years. Fortunately, the collapse occurred slowly enough that most of the people working on the site were able to escape. The building was cast-in-place reinforced concrete flat-slab construction with a central elevator shaft core. This style of construction is popular for multistory buildings because it reduces the slab thickness and the overall height of the building. The flat slabs were 190 mm (7 1/2 in.) thick, except for some bays near the elevator core and at stairwells, which were 230 mm (9 in.) thick. This design made possible a story height of 2.7 m (9 ft) for most of the floors. Figure 5-5 shows the floor plan.

The building, located at 2000 Commonwealth Avenue, was designed to be 16 stories high with a mechanical room above a 1.5 m (5 ft) crawl space on the roof. The building was 55.1 m × 20.9 m (180 ft 10 in. × 68 ft 6 in.) in plan. The structure also had two levels of underground parking. A swimming pool, ancillary spaces, and one apartment were located on the 1st floor, and 132 apartments were on the 2nd through 16th floors. Originally, these apartments were to be rented, but the owners later decided to market them as condominiums.

At the time of collapse, construction was nearing completion. Brickwork was completed up to the 16th floor, and the building was mostly enclosed from the 2nd to the 15th floors. Plumbing, heating, and ventilating systems were being installed throughout various parts of the building. Work on interior apartment walls had also started on the lower floors. A temporary construction elevator was located at the south edge of the building to aid in transporting equipment to the different floors. It is estimated that 100 people were working in or around the building at the time of the failure.

The failure took place in three phases. These phases were punching shear failure in the main roof at column E5, collapse of the roof slab, and, finally, the progressive and general collapse of most of the structure. Figure 5-6 shows the extent of the collapse.

Lessons Learned

A commission of inquiry was appointed by the mayor of Boston and convened a week after the collapse. The commission made a number of important findings (Granger et al., 1971):
  • Not a single drawing found in the file carried an architect's or engineer's registration stamp. The structural engineer refused to provide the calculations supporting his design to the commission. No principal or employee of the general contractor held a Boston builder's license.

- Ownership of the project changed a number of times, with changes in architects and engineers. This situation added to the overall confusion and contributed to the irregularities cited above. Some of the key changes are discussed in King and Delatte (2004).
- The general contractor only had a single employee on site, and most subcontracts were issued directly by the owner to the subcontractors and bypassed the general contractor. At least seven subcontractors were involved.
- The structural concrete subcontract did not require any inspection or cold weather protection of the work, although the designer had specified these measures. There was no evidence of any inspection of the work by an architect or engineer, although the project specifications required this.
- The concrete materials and quality control were poor.
- The triggering mechanism of the collapse was punching shear at the roof slab around column E5, probably preceded by flexural yielding of the roof slab adjacent to the east face of the elevator core.
- The construction did not conform to design documents and the construction procedures and materials were deficient. The design documents specified a 28-day strength of 20 MPa (3,000 lb/in.^{2}). At the time of the failure, 47 days after casting, the concrete had yet to achieve the required 28-day strength. The most significant deficiencies were a lack of shoring under the roof slab and low-strength concrete.

References

Delatte, Norbert J. (2009). *Beyond Failure: Forensic Case Studies for Civil Engineers*, ASCE Press, Reston, VA, 133 – 134.

Granger, R. O., Peirce, J. W., Protze, H. G., Tobin, J. J., and Lally, F. J. (1971). *The Building Collapse at 2000 Commonwealth Avenue, Boston, Massachusetts, on January 25, 1971*, Report of the Mayor's Investigating Commission, The City of Boston, MA.

King, S., and Delatte, N. J. (2004). "Collapse of 2000 Commonwealth Avenue: Punching Shear Case Study," *Journal of Performance of Constructed Facilities*, 18(1), 54–61.

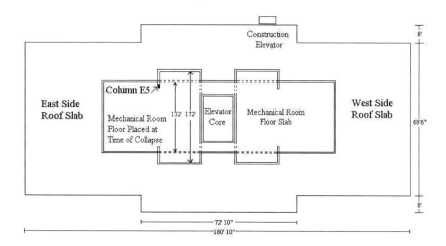

Figure 5-5. 2000 Commonwealth Avenue building floor plan.
Source: King and Delatte (2004), ©ASCE

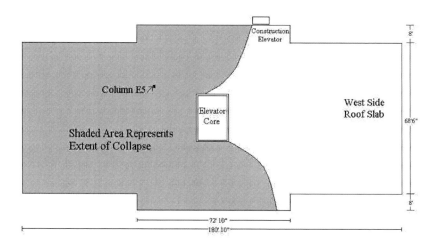

Figure 5-6. 2000 Commonwealth Avenue Building Extent of Collapse.
Source: King and Delatte (2004), ©ASCE

THE SKYLINE PLAZA APARTMENT BUILDING (BAILEY'S CROSSROADS)
(1973)

The Skyline Plaza complex located in Fairfax County, Virginia was a $200 million residential – commercial complex that included eight apartment buildings, six office buildings, a hotel and shops. Two of the apartment buildings had been completed with another pair under construction when one of the buildings under construction collapsed. Fourteen construction workers were killed, and thirty-five more were injured.

The apartment building that collapsed was to consist of 26 stories plus a penthouse and a four story basement. A parking structure was attached. The structure was of reinforced concrete flat plate construction with 20 cm (8 in.) concrete floor slabs and a typical story height of 2.7 m (9 ft).

On Friday, March 2, 1973, at approximately 2:30 p.m. a portion of one of the apartment buildings under construction collapsed. The collapse started with a section of the 23rd floor slab, directly under the 24th floor slab being cast, and proceeded vertically the full height of the building including the basement levels. Also destroyed in the collapse was the adjacent post-tensioned reinforced concrete parking facility (Figure 5-7).

Investigators agreed that the concrete had not gained adequate strength to carry the loads that were imposed on it during the construction process. No deficiencies were found in the original design. Engineers involved in the investigation concluded that the most probable triggering cause was punching at the 22nd floor columns.

Immediately after the incident, an inspection team from the Occupational Safety and Health Administration (OSHA) began an investigation of the site. A detailed investigation involving OSHA and the Center for Building Technology of the National Bureau of Standards (NBS) followed. While the cause of the collapse was directly related to poorly managed construction processes, a federal jury found the architect and consulting engineer guilty of negligence indicating that the building code required that the designers had a responsibility to inspect the work and warn those involved of any unsafe conditions.

Lessons Learned

Skyline Plaza was one of a series of progressive collapse failures that led to the implementation of special inspection sections in the building codes and requirements for review of progressive collapse issues during the design phase of structural projects. The investigation revealed serious violations of specified construction requirements and standard practices, including non-compliance with OSHA construction standards as follows:
  - Violation of requirements to fully-shore the two floors beneath the floor being cast.

- Failure to allow proper curing time before removal of shoring (specifically premature removal of the 22nd story forms).
- Failure to prepare or test field cured concrete specimens.
- Use of damaged and/or out of plumb shoring.
- Failure of inspections to note or correct violations.
- Improper "climbing" crane installation.

References

Breen, J. E., ed. (1975). *Summary Report – Research Workshop on Progressive Collapse of Building Structures*, Sponsored by the National Science Foundation, the National Bureau of Standards, and the U.S. Department of Housing and Urban Development.

Carino, N. J., Woodward, K. A., Leyendecker, E. V., and Fattal, S. G. (1983). "Review of the Skyline Plaza Collapse," *Concrete International Design and Construction*, 5(7) 45-42.

Civil Engineering (CE) (1975). "Bailey's Crossroads Synopsis," November, 59 – 61.

Cohen, E., Burns, N., and Meenen, A. (1974). *Skyline Towers Parking Structure Investigation*, Prestressed Concrete Institute.

Delatte, Norbert J. (2009). *Beyond Failure: Forensic Case Studies for Civil Engineers*, ASCE Press, Reston, VA, 144-149.

Dixon, D. E., and Smith, J. R. (1980). "Skyline Plaza North (Building A-4) A Case Study," *ASTM Special Technical Publication 702, Full-Scale Load Test of Structures*, Philadelphia, PA, 182-199.

Feld, J. and Carper, Kenneth L. (1997). *Construction Failure*, 2nd Ed., John Wiley & Sons, Inc., New York, NY, 242 – 243.

Leyendecker, E. V., and Fattal, S. G. (1977). *Investigation of the Skyline Plaza Collapse in Fairfax County, Virginia*, National Bureau of Standards, Building Science Series Number 94.

Ross, S. S. (1984). *Construction Disasters: Design Failures, Causes and Prevention*, McGraw-Hill.

Schlager, N. ed. (1994). "Skyline Plaza Collapse, Bailey's Crossroads, Virginia – 1973," *When Technology Fails*, Gale Research Inc., Detroit, MI, 277–282.

Schousboe, I. (1976). "Bailey's Crossroads Collapse Reviewed," *J. of the Construction Division*, Proceedings of the American Society of Civil Engineers, June, 365-378.

Figure 5-7. Overview of the Skyline Plaza Collapse at Bailey's Crossroads.
Source: National Bureau of Standards / National Institute of Standards and Technology

HARTFORD CIVIC CENTER COLISEUM
(1978)

Completed in 1973, the $75 million civic center in Hartford, Connecticut housed a coliseum, retail shops, and convention space. The coliseum roof collapsed before dawn on January 18, 1978 after a snowstorm and just hours after the last fans left a well-attended basketball game.

The structure consisted of a 10,000 sq. m (108,000 sq. ft) space truss roof, which was near record size for its time. The 6.5 m (21 ft) deep space truss had a plan dimension of 110 m by 91 m (360 ft by 300 ft), with clear spans of 64 to 82 m (210 to 270 ft) between the four pylon supports. The space truss consisted of 9.14 m x 9.14 m (30 ft x 30 ft) inverted pyramid modules with the top and bottom chords offset from each other by a half module in each direction (Figure 5-8). The chords and the main diagonals were each composed of four angles, arranged in a cruciform shape. Secondary intermediate members were attached to the midpoints of the chords and the main diagonals for purposes of bracing the main members. The roof framing system was elevated above the space truss, and was supported by short posts of varying length that landed on the main panel points, and some intermediate panel points. The space truss had been assembled on the ground and lifted into place. Computer methods were used for the original design and analysis.

During the night prior to the collapse a snowstorm deposited 15 psf to 18 psf of snow, as later estimated by the U.S. Army Cold Regions Research and Engineering Laboratory. The roof began to sag, and top chords of the truss buckled, redistributing the load to other members. These members were not able to carry the increased forces. As a result the roof folded and ultimately collapsed, dropping 25 m (83 ft) to the 12,000 seat arena below.

Lessons Learned

Investigation of the failure indicated several design deficiencies:
- Although the design was based on the assumption that the compression members were fully braced at midlength, they in fact were not. The interior members were only partially braced and the perimeter members, which were only braced about a single axis, were effectively unbraced over their entire 9.14 m (30 ft) length (Figure 5-9 and 5-10). The compression members were therefore much weaker in buckling than had been assumed by the designers.
- The use of the struts without cross-bracing did not provide diaphragm action for bracing the top chords.
- The roof dead loads were also seriously underestimated. A dead load of 40 psf was used by the designers, whereas the actual dead load was about 53.6 psf. This resulted in the total dead being underestimated by about 20%.

In addition, the investigation revealed that excessive deflection occurred during the construction phase, significantly exceeding the design deflection. This deflection was thought to indicate a potential problem. However, when this was relayed to the engineer of record, after reviewing the information, he dismissed it.

This failure has been appropriately referred to as a "computer-aided catastrophe." It illustrates the false sense of security offered by the data generated by computer analysis. This case serves as a lesson that computer software is a tool, and not a substitute for sound engineering experience and judgment. Relying only on computers for the design of increasingly complex structures, where the reasonableness and thoroughness of the analysis cannot be ascertained, invites failure.

References

Carper, Ken (2001). *Why Buildings Fail*, National Council of Architectural Registration Boards, Washington, D.C., 38 – 40.

Committee to Investigate the Coliseum Roof Failure, "Report of Committee to Investigate the Coliseum Roof Failure," City of Hartford, Conn., July 13, 1978.

Delatte, Norbert J. (2009). *Beyond Failure: Forensic Case Studies for Civil Engineers*, ASCE Press, Reston, VA, 174-184.

Engineering News-Record (ENR). (1978a). "Space Frame Roofs Collapse Following Heavy Snowfalls," *ENR*, January 26, 8.

ENR. (1978b). "Probe Closes in on Why Space Frame Roof Failed," *ENR*, March 16, 13.

ENR. (1978c). "Design Fault Suspected in Hartford Failure," *ENR*, March 30, 3.

ENR. (1978d). "Design Flaws Collapsed Steel Frame Roof," *ENR*, April 6, 9.

ENR. (1978e). "Someone Should Have Sounded Alarm," *ENR*, April 6, 88.

ENR. (1978f). "Space Truss, Not Space Frame," *ENR*, June 22, 29.

ENR. (1978g). "Collapsed Space Truss Roof had a Combination of Flaws," *ENR*, June 22, 36.

ENR. (1978h). "Collapsed Roof Design Defended," *ENR*, June 29, 13.

ENR. (1978i). "$12.3-Million Settlement in Roof Collapse," *ENR*, July 6, 3.

ENR. (1978j). "Design, Inspection Blamed in Roof Collapse," *ENR*, July 27, 15.

ENR. (1978k). "Hartford Coliseum Suit," *ENR*, November 30, 10.

ENR. (1978l). "On Collapse Report Authorship," *ENR*, September 7, 3.

ENR. (1978m). "Hartford: Designer's Duties," *ENR*, September 21, 21.

Lev Zetlin Associates, Inc., "Report of the Engineering Investigation Concerning the Causes of the Collapse of the Hartford Coliseum Space Truss Roof of January 18, 1978," New York, NY 1978

Levy, M. and Salvadori, M. (1992). *Why Buildings Fall Down*, Norton and Co., New York, NY, 68 – 75.

Martin, R., and Delatte, N. (2001). "Another Look at the Hartford Civic Center Coliseum Collapse," *Journal of Performance of Constructed Facilities*, 15(1), 31-36

Petroski, Henry (1985). *To Engineer Is Human*, St. Martins Press, New York, NY.

Redfield, R.K., Tobiasson, W.N., and Colbeck, S.C., "Estimated Snow, Ice, and Rain Load Prior to the Collapse of the Hartford Civic Center Arena Roof," U.S. Army Cold Regions Research and Engineering Laboratory, Hanover, NH, April 1979, Special Report 79-9.

Wearne, Phillip (2000). *Collapse: When Buildings Fall Down*, TV Books, L.L.C., New York, NY, 17–25.

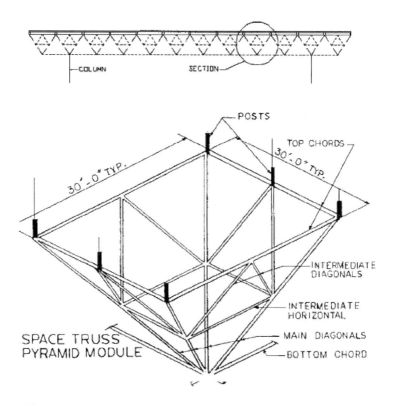

Figure 5-8. Elevation and detail of Hartford Civic Center roof.

Source: Drawing by Lev Zetlin Associates; reproduced with permission from Thornton Tomasetti.

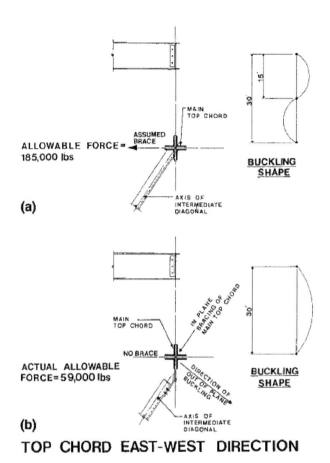

TOP CHORD EAST-WEST DIRECTION

Figure 5-9. Effect of assumed (hard) versus actual (spring) bracing condition.
Source: Drawing by Lev Zetlin Associates; reproduced with permission from Thornton Tomasetti.

	Connection A	Connection B	Connection C	Connection D
Original Design	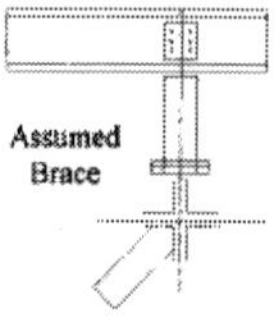 Assumed Brace Allowable force: **710 MN** (160,000 lb.) Allowable moment: 0	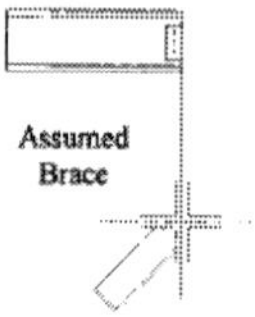 Assumed Brace Allowable force: **821 MN** (185,000 lb.)	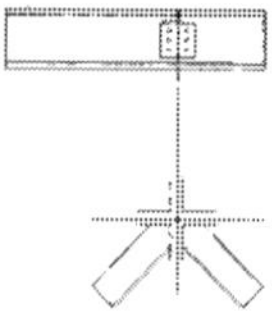 Allowable force: **2.78 GN** (625,000 lb.)	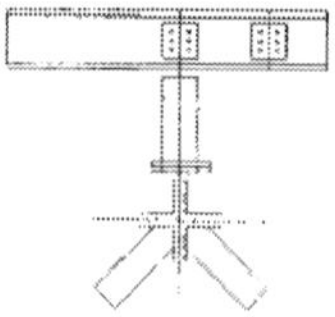 Allowable force: 2.51 GN (565, 000 lb.)
As Built	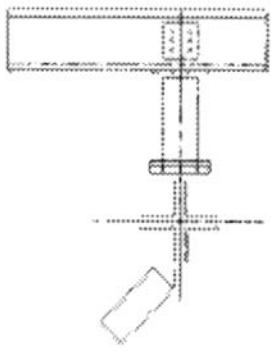 Allowable force: **68.6 MN** (15,440 lb.) Allowable moment: 12,800 N-m (9,490 lb.-ft.)	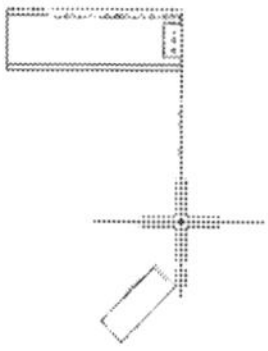 Allowable force: **262 MN** (59,000 lb.)	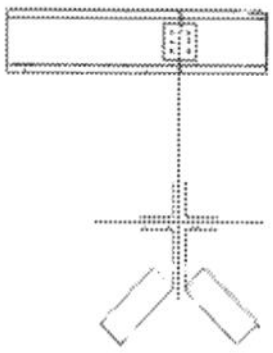 Allowable force: **1.61 GN** (363,000 lb.)	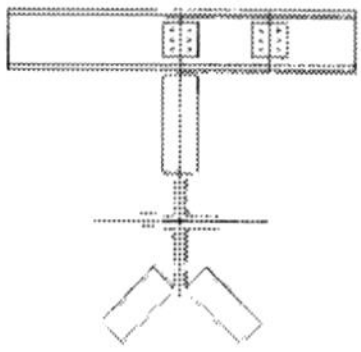 Allowable force: 2.51 GN (565, 000 lb.)

*Drawings by Rachel Martin based on information from ENR, April 6, 1978

Figure 5-10. Original and As Built Connection Designs.
Source: Martin and Delatte (2001), ©ASCE

IMPERIAL COUNTY SERVICES BUILDING
(1979)

The six story reinforced concrete Imperial County Services Building was only eight years old when it was crippled by the 1979 Imperial Valley earthquake, which had a Richter magnitude of 6.6. The structure comprised a five story box-like, relatively rigid, upper portion with an approximately 930 sq. m (10,000 sq. ft) footprint that was supported on columns and irregularly placed shear walls at the first floor level. The columns at the east end of the building experienced severe damage with bursting of the rebar ties and concrete crushing resulting in significant shortening (Figure 5-11). The building was judged unrepairable and was subsequently demolished.

The exceptional circumstance of this building was the presence of a 13-transducer array of strong motion recorders which provided, for the first time ever, a set of data recording the behavior of a building which experienced major structural damage. This fact, coupled with the availability of measurements made of the significant dynamic characteristics prior to the earthquake, allowed detailed analyses of the failure sequence.

Lessons Learned

The fact that architectural configuration has an important influence on the ability of a building to resist earthquakes was confirmed by the behavior of the Imperial County Services Building in the 1979 earthquake.

References

Earthquake Engineering Research Institute (EERI) (1980). *Reconnaissance Report, Imperial County, California, Earthquake of October 15, 1979*, EERI.

Shepherd, R., and Plunkett, A.W. (1983). "Analysis of the Imperial County Services Building," *J. of Structural Engineering Division*, 109 (STI), 1711-1726.

Figure 5-11. Column of Imperial County Services Building.
Source: C.E. Johnson, USGS.

KEMPER MEMORIAL ARENA ROOF
(1979)

A severe wind and rain storm that swept through the Kansas City, Missouri area on June 4, 1979 was blamed for the collapse of a significant portion of the roof of the Crosby Kemper Memorial Arena (Figure 5-12). Fortunately, at the time of collapse, 7:10 p.m., the 17,600-seat arena was occupied by only two maintenance personnel, neither of whom was injured. Just two days earlier, the arena had been packed with 13,500 people for a major event.

The roof structure was severely damaged as 4000 sq. m (43,000 sq. ft) of the 12,000 sq. m (130,000 sq. ft) roof had to be replaced along with some repairs to the hanger system that connected the general roof framing to the large external exposed architectural space frame. The primary space frames did not collapse and were relatively undamaged as a result of the roof failure.

The arena is a reinforced concrete seating and service structure, enclosed by metal wall panels and a roof hung from external structural steel pipe space frames. Three individual triangulated space frames spaced at 30m (99 ft) on center spanned 97 m (324 ft) and provide support for the secondary roof of steel plane trusses at 42 different panel points. Large diameter steel pipe members comprise the space frames. Pipe hangers were used to connect the primary and secondary roof systems. The hangers were fastened to the space frames by a welded gusset plate with a pin connection, and to the top chords of the secondary roof trusses by a steel baseplate. Four A490 bolts loaded vertically in tension connected the pipe hangar baseplate to the secondary steel trusses.

During the five years the building was occupied wind loading and unloading induced rocking action on the bolts of the connection. These dynamic effects were determined to be the most probable cause of the collapse eventually leading to fatigue failure of the A-490 bolts. The Seventh Edition of the Manual of Steel Construction restricted the use of A490 bolts in tension to static applications only. Subsequent information and test data have led to the lifting of this restriction starting with the 2000 Specification for Structural Joints Using ASTM A325 or A490 Bolts.

Contributing factors to the bolt failure was the manner in which the connection was configured which led to prying action and bolts that were not fully tightened due to the presence of a flexible Micarta plate between the pipe hangar and the roof trusses. The lack of redundancy in the basic design, a common problem in modern long span structures, was also cited as a deficiency. In this case, fatigue failure in a single bolt was able to bring about a progressive collapse of the entire center section of the roof and cause significant damage to other areas.

Lessons Learned

Several dramatic structural failures occurred within the span of a couple of years. These included the Hartford Connecticut Civic Center Coliseum roof (1978), the Kansas City Kemper Arena (1979), and the Hyatt Regency Hotel walkways (1981).

Taken together, these failures had a sobering effect on the design and construction industry. Attention was focused on the need to provide greater structural integrity and redundancy in the design of structures. In particular, the importance of connection design was emphasized. Detailing and execution of connections, including provisions for their adequate review during the shop drawing phase, has been the subject of significant discussion since these failures occurred. In addition, this failure and others have contributed to the understanding of wind forces on buildings and the evolution of improved wind design standards to consider dynamic effects and pressure concentrations.

References

Carper,   Ken (2001). *Why Buildings Fail*, National Council of Architectural Registration Boards (www.ncarb.org), Washington, D.C.

CE. (1981). "Kemper Arena Roof Collapse and Repair," March, 68-69.

Delatte, Norbert J. (2009). *Beyond Failure: Forensic Case Studies for Civil Engineers*, ASCE Press, Reston, VA, 124-126.

Dorris, Virginia Kent (1994). "Kemper Arena Roof Collapse," *When Technology Fails*, ed. Neil Schlager, Gale Research, Inc. Detroit, MI, 295 – 299.

ENR. (1979a). "Failed Bolt Connections Bring Down Arena Roof," *ENR*, June 14, 10-12.

ENR. (1979b). "Clean-up Begins as Kemper Designer Sues for Access," *ENR*, June 28, 13.

ENR. (1980). "Kemper Arena Reopens with New Roof Connections," *ENR*, March 13, 18.

Feld, Jacob and Carper, Kenneth L. (1997). <u>Construction Failure</u>, John Wiley & Sons, Inc., New York, NY.

Goldberger, Paul (1979). "Kansas City Arena Loses Roof in Storm," *New York Times*, June 6, A1, A25.

Gustafson, Kurt (2004). "ASTM A490 Bolts and Fatigue Loads," *Modern Steel Construction*, American Institute of Steel Construction, Chicago, IL.

Kaminetzky, Dov (1991). *Design and Construction Failures. Lessons From Forensic Investigations*, McGraw-Hill Book Company, New York, NY.

Levy, Matthys, and Salvadori, Mario (1992). *Why Buildings Fall Down*, W. W. Norton, New York, NY, 57-67.

Ross, Steven (1984). *Construction Disasters: Design Failures, Causes, and Prevention*, McGraw-Hill, New York, NY.

Stratta, J. L. (1979). *Report of the Kemper Arena Roof Collapse of June 4, 1979, Kansas City Missouri*, James L. Stratta, Consulting Engineer, Menlo Park, CA.

Snoonian, Deborah (2000). "Sleuthing Out Building Failures," *Architectural Record*, Aug., 163-170.

Wearne, Phillip (2000). *Collapse: When Buildings Fall Down*, TV Books, L.L.C., New York, NY, 17-36.

Figure 5-12. Kemper roof debris litters seats below shortly after the collapse.
Source: Courtesy of Wiss, Janney, Elstner Associates, Inc.

BINGHAMTON STATE OFFICE BUILDING
(1981)

In February of 1981, an electrical transformer fire broke out in the basement of the 18 story State Office Building in Binghamton, New York. The fire started at 5:30 a.m. and lasted for a total of only about 30 minutes. The result, however, was one of the worst cases of chemical contamination of a building in US history. This incident is sometimes referred to as the first "indoor environmental disaster."

A fault in the 480 V secondary switch gear in the basement mechanical room of the office building was blamed for starting the fire. The intense heat that was generated by the fire cracked a bushing on a nearby askeral filled transformer. About 180 gallons of this insulating fluid, known to contain PCBs (polychlorinated biphenyls) escaped from the transformer which in turn vaporized and mixed with smoke and soot from the fire. When the fire alarm was triggered, hatches in the roof above the stairwells automatically opened. When the firefighters arrived and opened the door to the mechanical room, the contaminated smoke and soot was drawn up the stairwells in a chimney-like effect spreading the toxic contaminants to the ventilation systems which in turn dispersed toxic soot containing PCBs, dioxin and dibenzofurans (furnans) throughout the building.

The immediate problems of the fire were compounded by misunderstandings relative to the severity of the contamination and mismanagement of the overall cleanup process. The building had been constructed in 1972 at a cost of $17 million, but the cleanup took nearly 14 years and cost $53 million. The disaster also spawned a legal process that lasted over 20 years. Litigation ended in 2004 with a $7.2 million settlement between the various claimants and the two companies who manufactured the electrical equipment and transformer coolant respectively.

A similar fire related incident occurred in December of 1991 at the State University of New York at New Paltz, contaminating several buildings and resulting in a first phase clean-up operation that lasted over three years. In June of 1985, a transformer located in the basement of the New Mexico State Highway Department overheated and released an oily mist containing PCBs in an askarel fluid which extensively contaminated the three-story building, partially compounded by a clean-up effort that took place before the PCB contamination was identified.

Lessons Learned

PCBs, such a those contained in older transformers, are considered to be toxic substances that may result in serious health concerns. It has been found that when burned, PCBs generate by-products which include polychlorinated dibenzobioxin (Dioxin) and polychlorinated dibenzofurans that are even more toxic than the PCBs themselves. Transformers containing PCBs were manufactured between the years 1929 and 1977 and were subsequently banned for use in manufacturing transformers and some other electrical devices. The EPA now regulates the use, location, storage and disposal of transformers containing PCBs.

In addition, as a result of this failure and a number of other high profile fires of various types, the industry, including the major building codes, are placing a much greater emphasis on smoke control and smoke evacuation in buildings.

References

DesRosiers, (1984). "PCB's, PCDF's, and PCDD's resulting from transformer/ capacitor fires: an overview," *Proceedings: 1983 PCB seminar, EPRI EL-3581, project 2028*, Addis G. and Komai R.Y., eds., Palo Alto, California: Electric Power Research Institute, 6-41 to 6-57.

Environmental Protection Agency (EPA) (2007). *Polychlorinated Biphenyls (PCBs)*, http://www.epa.gov/reg3wcmd/ts_pcbs.htm (accessed on May 14, 2007).

National Academy of Sciences. (2005). *Reopening Public Facilities After a Biological Attack: A Decision-Making Framework*, National Academy of Sciences, 193-194, http://www.nap.edu/openbook/0309096618/html/193.html (accessed May 18, 2006).

O'Keefe, P.W., Silkworth, J.B., Gurthy, J.F., et al. (1985). "Chemical and biological investigations of a transformer accident at Binghamton NY," *Environmental Health Perspect.*, 60:201-9.

Schecter, A. (1983). "Contamination of an Office Building in Binghampton, New York by PCBs, Dioxins, Furans and Biphenylenes After an Electrical Panel and Electrical Incident," *Chemosphere*, 12(4/5), 669-680.

Schecter, A. (1986). "The Binghamton State Office Building PCB, Dioxin and Dibenzofuran Electrical Transformer Incident: 1981-1986," *Chemosphere*, 15(9-12), 1273-1280.

Simonson, Mark (2006). "Office building becomes toxic tower," *The Daily Star*, http://www.thedailystar.com/opinion/columns/simonson/2006/01/simonson01 30.html (accessed May 18, 2006).

Vogt, Barbara Muller, and Sorensen, John H. (2002). *How Clean is Safe? Improving the Effectiveness of Decontamination of Structures and People Following Chemical and Biological Incidents*, U.S. Department of Energy Chemical and Biological National Security Program, Oak Ridge National Laboratory, 14-15.

HYATT REGENCY HOTEL PEDESTRIAN WALKWAYS
(1981)

On July 17, 1981 two suspended walkways in the atrium area of the Hyatt Regency Hotel in Kansas City, Missouri collapsed suddenly. The failure caused 114 deaths and 185 injuries. Because of the large number of casualties, this failure had a profound and lasting impact on the design and construction industry.

The hotel had been in service for a full year prior to the collapse. The walkways, suspended by tension rods from the atrium roof structure, were arranged such that the second floor walkway was suspended directly below the fourth floor walkway. A separate walkway at the third floor level was not involved in the collapse.

Failure of the walkways was due to a combination of procedural errors and technical causes. The technical cause of the failure was quickly established and later confirmed by numerous investigators. A deficient connection at the juncture of suspension rods and a box beam built up of welded channels failed, causing the fourth and second floor walkways to collapse to the floor in a pile of debris that weighed 570 KN (64 tons) (Figure 5-13). From a procedural standpoint, the original single rod connection had been modified to a two rod support during the construction process and the connection was never designed, neither by the engineer of record nor by the fabricator's engineer. The detail also anchored the rods along the welded seam of two standard steel channels facing inward to form a box-shape - a weak point in the connection (Figure 5-14).

Structural tests conducted at the National Bureau of Standards clearly demonstrated that the connections as originally shown on the design drawings were not capable of supporting the gravity load required by the relevant building code. This deficiency was compounded by the support rod change made to the detail during construction, which doubled the load on the connection, making its failure inevitable.

There was very little argument within the industry relative to the technical explanation of the failure. However, discussion about the chain of events that permitted this collapse to occur was extremely significant and continues to this day. Questions regarding deficiencies in the project delivery system were the focus of landmark litigation and administrative hearing decisions. These questions were quite basic: Who designed the original connection? Was the original connection buildable? Who initiated the change to the connection during the construction phase and why? Who approved it? What is the meaning of "shop drawing review?" This failure led directly to ongoing activities aimed at improving quality assurance and quality control in the design and construction process.

Lessons Learned

In terms of human suffering, the Hyatt Regency Hotel walkway collapse was one of the most significant tragedies to ever confront the construction industry. The technical cause of the failure was easy to understand. However, there were many lessons for design and construction professionals. The important lessons involved procedural issues. Clearly, there is a need for all parties to understand their responsibilities and to perform their assignments competently. The structural engineer's responsibility for overall structural integrity, including the performance of connections was firmly established in this case. This failure also reinforced the need for practices such as project peer review and constructability checks.

References

Carpenter, Carl H. (2001). Discussion, "Special Section: Legacy of the Kansas City Hyatt Tragedy: A 20-Year Retrospect Insight and Review," *Journal of Performance of Constructed Facilities*, 15(4), 154.

Dallaire, G. and Robison, R. (1983). "Structural Steel Details: Is Responsibility the Problem?," *Civil Engineering*, October, 51-55.

Delatte, Norbert J. (2009). *Beyond Failure: Forensic Case Studies for Civil Engineers*, ASCE Press, Reston, VA, 8-25.

Gillum, J. D. (2000). "The Engineer of Record and Design Responsibility," *Journal of Performance of Constructed Facilities*, 12(2), 67-70.

Leonards, G. A. (1983). "FORUM: Collapse of the Hyatt Regency Walkways: Implications," *Civil Engineering*, March, 6.

Luth, G. P. (2000). "Chronology and Context of the Hyatt Regency Collapse," *Journal of Performance of Constructed Facilities*, 12(2), 51-61.

Moncarz, P. D. and Taylor, R. K. (2000). "Engineering Process Failure – Hyatt Walkway Collapse," *Journal of Performance of Constructed Facilities*, 12(2), 46-50.

NBS (1982). Investigation of the Kansas City Hyatt Regency Walkways Collapse, NBSIR 82-2465, National Bureau of Standards (National Institute of Standards and Technology), Department of Commerce, Washington, DC.

Paret, Terrence F., Moncraz, Piotr D., and Taylor, Robert K. (2001). Discussion, "Engineering Process Failure – Hyatt Walkway Collapse," *Journal of Performance of Constructed Facilities*, 15(4), 154.

Pfang, E. O. and Marshall, R. (1982). "Collapse of the Kansas City Hyatt Regency Walkways," *Civil Engineering*, July, 65-68.

Pfatteicher, S. K. A. (2000). "The Hyatt Horror: Failure and Responsibility in American Engineering," *Journal of Performance of Constructed Facilities*, 12(2), 62-66.

Robison, R. (1984). "Structural Steel Details: Comments on Divided Responsibility," *Civil Engineering*, March, 58-60.

Rubin, R. A. and Banick, L. A. (1987). "The Hyatt Regency Decision: One View," *Journal of Performance of Constructed Facilities*, 1(3), 161-167.

Figure 5-13. Second and Fourth Floor Skywalks collapsed in lobby.
Source: Courtesy of Wiss, Janney, Elstner Associates, Inc.

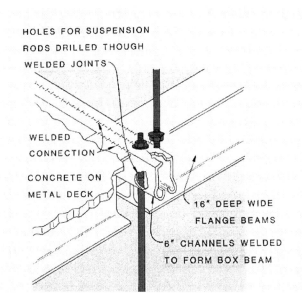

Figure 5-14. Detail of deficient connection and collapse mechanism.
Source: Courtesy of Wiss, Janney, Elstner Associates, Inc.

PINO SUAREZ BUILDING
(1985)

The Pino Suarez was constructed in Mexico City in the early 1970s and consisted of five high- rise steel buildings supported on a two-level reinforced subway station that acted as a rigid foundation common to all five buildings. One of the functions of the complex was to act as a counterweight against uplift forces caused by the expansive soils which occurred as the consequence of the large excavation for the subway station. In the middle of the complex were three identical 21-story buildings each two bays wide and four bays deep, aligned and parallel, with their narrow sides facing east/west. Directly adjacent to, but north and south of the 21 story buildings were two identical 14 story towers. These two buildings were positioned with their narrow faces directly north and south respectively relative to the longitudinal sides of the 21 story buildings. The structural framing of the 21-story buildings consisted of truss beams connected to hollow box columns with moment resisting connections. In addition a bracing system was employed that consisted of two cross-braced frames in the transverse direction and one V-braced frame in the exterior longitudinal plane.

As a result of the September 19, 1985 Mexico City earthquake, the complex suffered severe structural failure. One of the three 21-story steel towers collapsed onto an adjacent 14 story tower, destroying it also. The performance of the other two 21-story towers that were severely damaged but did not collapse provided invaluable information with respect to the progression of failure. Computer simulations confirmed the progression of failure to a reasonable degree of engineering certainty. It was inferred that plastic hinges first developed at the girder ends leading to yielding and plate buckling of the two fourth story columns located on the south side (Figure 5-15). Columns failing in this way lost most of their gravity load carrying capacity and suffered shortening. This led to buckling of the X-brace framing and the redistribution of forces throughout the structure that eventually caused the failure of the other four story columns and the collapse of the tower.

Lessons Learned

The Pino Suarez complex failure was caused by a design flaw that was not recognized in the then current seismic design codes. After the failure, Mexico updated its seismic design code recognizing the need to concentrate energy dissipation in less critical "fuse" elements, and protect the main columns and other key structural elements. Although member yielding provides a great energy dissipation capability, its effects should be explicitly considered in design. Sufficient connection strength is necessary and yielding should be confined primarily to beams and bracing not to columns and joints. Finally, the importance of structural redundancy was reemphasized by the failure of the Pino Suarez complex.

References

Hanson, R.D. and Martin, H.W. (1987). "Performance of Steel Structures in the September 19 & 20,1985 Mexico Earthquakes," *Earthquake Spectra*, 3(2), 329-345.

Osteraas, J. and Krawinkler, M. (1989). "The Mexico Earthquake of September 19, 1985 - Behavior of Steel Buildings," *Earthquake Spectra*, 5(1), 51-87.

Figure 5-15. Pino Suarez Buckled Column.
Source: E.V. Leyendecker, USGS.

L'AMBIANCE PLAZA
(1987)

On April 23, 1987, at approximately 1:30 p.m., the structural frame and slabs of what was to be a 16-story apartment building in Bridgeport, Connecticut collapsed during construction killing 28 construction workers. This was the largest loss of life in a construction accident in the United States since 51 workers were killed in the collapse of a reinforced concrete cooling tower under construction at Willow Island, West Virginia in 1978.

The L'Ambiance Plaza project comprised two rectangular towers, each with 13 levels of apartments and three levels of parking (Figure 5-16). The towers were being constructed using the lift-slab method. Floor and roof slabs were two-way, unbonded, post-tensioned flat plates. They were all cast at the ground level, post-tensioned and then lifted into place (Figure 5-17). Steel columns supported the floor slabs, which were lifted in stages by hydraulic jacks, threaded rods, and welded steel shearhead collars placed in the concrete floor slab at each column (Figure 5-18).

The collapse investigation was complicated by a number of factors, such as the sheer size of the collapse, the unusual building system, and the unavoidable damage to evidence that occurred during the frantic rescue operation. Several triggering causes were hypothesized, most centered on deficiencies in the design and construction of the shearhead collars. One theory proposed that a lifting angle in the shearhead deformed, allowing the nut at the end of the lifting rod to slip out of the assembly. Another theory involved a rolling out of the wedges used to temporarily support slabs during the lifting operation (Figure 5-19). Deficient welds were also discovered in the shearheads. Other failure theories included improper design of the floor slab post-tensioning tendons, overall (global) instability, and foundation failure under one of the columns.

Several factors present in the design and construction sequence were identified as either triggering mechanisms or contributors to the magnitude of this catastrophe. These included the improper placement of some of the prestressing tendons and general overall instability of the frame during construction. Shear wall construction lagged behind the lifting operation, leaving the upper part of the structure unstable.

A remarkable and controversial mediated global settlement of claims resulting from this failure brought an end to forensic investigations in December 1988. There may never be complete general agreement on the technical causes of the failure, but one result of the early settlement was that the various theories were discussed thoroughly in the engineering literature, much sooner than has been the case for other collapses of this magnitude.

Lessons Learned

The tragedy of the L'Ambiance Plaza collapse gave impetus to efforts aimed at ensuring structural integrity during the construction phase of a project. This failure underlined the need to carefully evaluate details and sequencing specifically for each project and each site.   Much discussion also centered on the convoluted and fragmented project delivery system, in which responsibility for ultimate structural safety was confused by unclear relationships among the engineer of record, the lift-slab contractor, and the designer of the shearheads.

References

Carper, Ken (2001). *Why Buildings Fail,* National Council of Architectural Registration Boards, Washington, D.C., 66 – 71.

Culver, C. G. (2002). Discussion of "Another Look at the L'Ambiance Plaza Collapse," *Journal of Performance of Constructed Facilities*, 16(1), 42.

Cuoco, D., D. Peraza and T.Z. Scarangello (1992). "Investigation of the L'Ambiance Plaza Building Collapse," *Journal of Performance of Constructed Facilities*, 6(4), 211-231.

Delatte, Norbert J. (2009). *Beyond Failure: Forensic Case Studies for Civil Engineers*, ASCE Press, 107-121.

Felsen, M.D. (1989). "Mediation that Worked: Role of OSHA in L'Ambiance Plaza Settlement," *Journal of Performance of Constructed Facilities*, 3(4), 212-217.

Heger, F.J. (1991). "Public Safety Issues in Collapse of L'Ambiance Plaza," *Journal of Performance of Constructed Facilities*, 5(2), 92-112.

Martin, R., and Delatte, N. (2000). "Another Look at the L'Ambiance Plaza Collapse," *Journal of Performance of Constructed Facilities*, 14(4), 160-165.

McGuire, W. (1992). "Comments on L'Ambiance Plaza Lifting Collar/Shearheads," *Journal of Performance of Constructed Facilities*, 6(2), 78-95.

Moncarz, P.D., Hooley, R., Osteraas, J.D. and Lahnert, R.J. (1992). "Analysis of Stability of L'Ambiance Plaza Lift-Slab Towers," *Journal of Performance of Constructed Facilities*, 6(4), 232-245.

Poston, R., Feldmann, G.C. and Suarez, M.G. (1991). "Evaluation of L'Ambiance Plaza Post-tensioned Floor Slabs," *Journal of Performance of Constructed Facilities*, 5(2), 75-91.

Scribner, C.F. and Culver, C.G. (1988). "Investigation of the Collapse of L'Ambiance Plaza," *Journal of Performance of Constructed Facilities*, 2(2), 58-79.

Zallen, R.M., and Peraza, D.P. (2004). *Engineering Considerations for Lift-Slab Construction,* ASCE, Reston, VA.

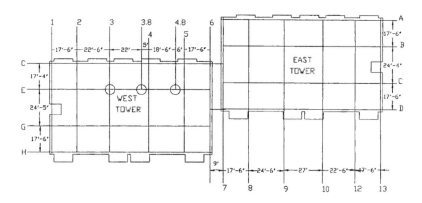

Figure 5-16. Floor plan of L'Ambiance Plaza.
Source: Martin and Delatte (2000), ©ASCE

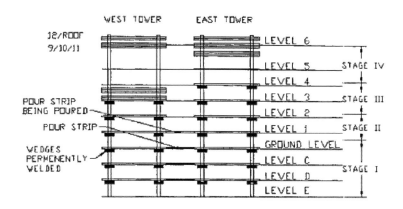

Figure 5-17. Elevation of L'Ambiance Plaza just before collapse.
Source: Martin and Delatte (2000), ©ASCE

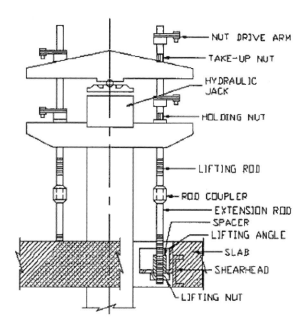

Figure 5-18. L'Ambiance Plaza Lifting assembly.
Source: Martin and Delatte (2000), ©ASCE

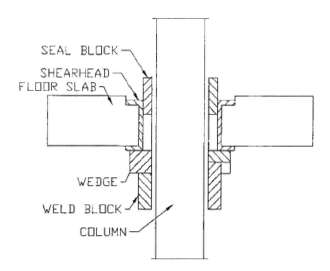

Figure 5-19. L'Ambiance Plaza Wedges.
Source: Martin and Delatte (2000), ©ASCE

BURNABY SUPERMARKET ROOFTOP PARKING DECK
(1988)

Part of the rooftop parking deck of a supermarket at the Station Square development in Burnaby, British Columbia, Canada, collapsed on opening day, during a Grand Opening ceremony on April 23, 1988. The ceremony was a special preview opening for neighborhood senior citizens. They had been directed to park on the 69 by 122 m (225 by 400 ft), one-story building, which was part of a regional community shopping center that included a hotel, apartments, retail space, theaters, offices and the flagship building for the Save-on-Foods retail chain. After a welcoming program, the 600 senior citizens began to shop, aided by 370 employees of the Save-on-Foods store.

About 15 minutes later a loud bang was heard, and water began to flow from a broken overhead fire sprinkler pipe. A photograph was taken of the broken pipe and a severely distorted beam-to-column connection. After this distortion and pipe rupture, the supermarket staff acted promptly and efficiently to clear people from the immediate area and then to begin evacuation of the entire store. Approximately 4-1/2 minutes later the roof in four bays collapsed into the shopping area causing considerable building damage along with destroying 20 automobiles. The collapsed area was 27 by 23 m (87 by 75 ft). No one was killed, but 21 people were injured.

A commissioner was appointed two weeks after the failure by the government of British Columbia to investigate the failure. The commissioner's report detailed the probable technical cause of the collapse and also reviewed the many procedural deficiencies that led to this failure. The recommendations included in the report continue to effect important revisions to standards of practice in Canada.

The 8,400 sq. m (90,000 sq ft) supermarket building was a single-story steel column and beam structure that used the cantilevered or Gerber beam system. Wide-flange steel beams passed over hollow steel tube columns, extending as cantilevers at each end. The beams supported open-web steel joists, which in turn supported a composite concrete and corrugated metal deck. The most probable technical cause of the failure was insufficient stability of the beam-to-column connection. There was no provision in the design for lateral support to the bottom flange of the beam, at a condition of bending moment that placed this flange in compression. This type of failure has occurred in the past, and the need to investigate this mode of failure has been well-established.

Lessons Learned

Of greater interest than the technical cause of this failure are the many procedural deficiencies in the project delivery system that permitted the design deficiency to go unrecognized. The commissioner's report cites numerous contributing procedural problems, including competitive bidding for design services, unclear assignment of responsibilities, inadequate involvement of designers during the construction phase, poorly-monitored changes during construction, incomplete peer review, and inadequate professional liability insurance. These procedural problems are certainly not unique to Canada, but are the same set of deficiencies that plague the construction industry in the United States. The commissioner's report contained 19 recommendations, including: 1) independent project peer review funded by increased permit fees; 2) special examinations for structural engineers and mandatory professional liability insurance; 3) the development by the provincial government of a manual that would clarify the responsibilities of all parties to the construction process; 4) a minimum fee schedule for design services; and 5) strengthening of certain steel industry guidelines and design manuals, particularly with respect to beam-column connection support requirements.

References

Galambos, T. V., ed. (1987). *Guide to Stability Design Criteria for Metal Structures*, 4th edition, John Wiley and Sons, New York, NY.

Government of British Columbia (1988). *Report of the Commissioner Inquiry - Station Square Development,* Victoria, B. C., Canada.

Jones, C. P. and Nathan, N.D. (1990). "Supermarket Roof Collapse in Burnaby, B.C., Canada," *Journal of Performance of Constructed Facilities*, 4(3), 142-160.

Baer, B. R. (1992). "Discussion: Supermarket Roof Collapse in Burnaby, B.C., Canada," *Journal of Performance of Constructed Facilities*, 6(1), 67-68.

Jones, C. P. and Nathan, N.D. (1992). "Closure To Discussions: Supermarket Roof Collapse in Burnaby, B.C., Canada," *Journal of Performance of Constructed Facilities*, 6(1), 69-70.

NORTHRIDGE MEADOWS APARTMENTS
(1994)

A large proportion of the most severe damage to buildings in the January 17, 1994, Northridge earthquake in Southern California was experienced by condominium buildings. Several thousand of these structures were situated in the most severely shaken areas, and the largest single incidence of fatalities - 16 individuals - occurred in one such building, the Northridge Meadows Apartments.

The typical configuration of these buildings provides street level "tuck under" parking, either of timber framed carport configuration or comprising a reinforced concrete structural slab supported on perimeter walls and internal columns, above which is constructed two or three levels of stucco clad, timber framed, residential units.

A consistent pattern of seismic weaknesses was observed. These include overloading of the perimeter walls, incipient punching shear damage of column/slab connections, in-plane shear failures of stucco and drywall clad walls, cracking of vertical timber studs, splitting of bottom sill plates, shattering of lightweight concrete floor slabs, out-of-plane separation of stucco, and many connection detail deficiencies (Figure 5-20).

A particularly interesting aspect of the behavior of the condominium buildings in the Northridge earthquake is that many clearly experienced a seismic load demand that closely approximated the available capacity. Consequently the response observations provided an invaluable data base with which to assess the effectiveness of the design and construction practices that were followed.

In the case of the Northridge Meadows Apartments, the collapse of several blocks can be ascribed directly to the soft bottom story, as a consequence of insufficient shear resistance being provided at this level.

Lesson Learned

The style of multi-story, multi-family residential buildings typified by the Northridge Meadows Apartments proved hazardous in the moderate to strong earthquake shaking experienced on January 17, 1994. The frailties of the structural form were exposed. Stricter code provisions and more stringent inspection procedures to ensure improved quality control of the construction process will reduce the possibility of an event like this from happening in the future.

References

"Preliminary Report of the Seismological and Engineering Aspects of the January 17, 1992, Northridge Earthquake," (1994). Report UCB/EERC 94/01, Earthquake Engineering Research Center, University of California at Berkeley.

Hall, John. F. ed. (1994). *Northridge Earthquake, January 17, 1994; Preliminary Reconnaissance Report*, Earthquake Engineering Research Institute, Oakland, California.

Figure 5-20. Collapsed first floor of Northridge Meadows Apartment.

Source: J. Dewey, USGS.

CALIFORNIA STATE UNIVERSITY, NORTHRIDGE, OVIATT LIBRARY
(1994)

The Magnitude 6.4 January 17, 1994, Northridge earthquake in Southern California, although not a major event in terms of total energy release, was relatively shallow with its epicenter in a developed urban area. Consequently the local ground shaking was very intense, the damage to constructed facilities was severe and the economic loss was huge. First indications were that only minor damage had been sustained by steel-framed structures. One such building was the four story steel braced frame Oviatt Library building on the campus of California State University, Northridge. As is common with many flexible structures, architectural finishes were found to have been damaged but this was ascribed to the inherent incompatibility of the relatively flexible steel skeleton and the stiffer cladding. The observation that substantially greater, readily apparent, architectural damage occurred in some of the aftershocks resulted in a more thorough inspection being undertake. This revealed fractures in the steel column base plates and in various other welded connection in the steel framework.

Since steel framed structures comprised the seismic resisting system of choice for tall buildings, the established fact that this public building had not performed satisfactorily, coupled with unconfirmed reports of several unexpected failures of other steel frames, prompted a rigorous inspection of all the buildings falling into this category in the zone of strong to moderate ground shaking.

It was found that in more than 100 buildings brittle fractures had occurred in welded connections, including the moment resisting joints between beams and columns. Detailed analyses indicated that these buildings had been designed and constructed in conformity with normal industry standards.

The disturbing conclusion drawn from these observations was that structural engineers could no longer have confidence in the established procedures used to design and construct steel framed buildings in earthquake prone areas, notwithstanding that these buildings were in compliance with the appropriate codes. Additionally, of even greater concern in the short term, no consensus existed as to an acceptable procedure for repair of the damaged frames.

Faced with an exceptionally challenging compound technical problem, those concerned with the re-establishment of steel moment resistant frame buildings as a viable alternative amongst the several structural choices available to earthquake engineers, cooperated to devise and initiate a program with the focused objective of leading to the development of standards for the repair, retrofit and design of steel moment resisting frame buildings so that they will provide reliable, cost-effective performance in future earthquakes.

The experience of the directed program of professional practice development coupled with problem focused investigations is likely to be reviewed with great interest by those who may wish to consider the applicability of this approach to other engineering challenges.

Lessons Learned

The Northridge earthquake revealed unforeseen vulnerabilities in the seismic resisting system of choice for tall buildings, namely a steel frame with moment resisting beam-column.

However, as is so often the case with technological progress, the pressures to adopt new, supposedly more cost efficient, techniques outran the performance verification process necessitating major reconsideration of the recently accepted industry standards.

References

American Institute of Steel Construction (AISC) (1994). *Northridge Steel Update I*, AISC Inc., Chicago, IL

Bertero, V.V., Anderson, James, C. & Krawinkler, H. (1994). *Performance of Steel Building Structures during the Northridge Earthquake*, Report No. UCBIEERC-94/09, Earthquake Engineering Center, University of California, Berkeley.

City of Los Angeles (1994). *Repairs of Cracked Moment Frame Connections in Welded Steel Frame Structures and Requirements for SMRF-Welded Connections in New Buildings*, Plan Check Information.

California Seismic Safety Commission (CSSC) (1994). "Damage to Steel Frame Buildings by the Northridge Earthquake," Public Advisory issued by the CSSC, Sacramento, CA

Hall, John. F. ed. (1994). *Northridge Earthquake, January 17, 1994; Preliminary Reconnaissance Report*, Earthquake Engineering Research Institute, Oakland, CA

Modern Steel Construction (1994). "Localized Steel Damage," American Institute of Steel Construction (AISC) Inc., 34(4).

SAC (1994). *Program to Reduce Earthquake Hazards in Steel Moment Frame Structures*, formulated by the SAC Joint Venture Partnership, Structural Engineers Association of California, Sacramento, CA

SAC (1994). *Steel Moment Frame Connection, Advisory No. 1*, SAC Joint Venture Partners, Structural Engineers Association of California, Sacramento, CA

ALFRED P. MURRAH FEDERAL BUILDING
(1995)

On April 19, 1995, a truck bomb tore through the façade of the Murrah Federal Building in Oklahoma City, OK. The blast destroyed a significant portion of the structure, killing 169 people. This event was an example of what is known as a progressive collapse. A progressive collapse or disproportionate collapse occurs as an event that should be localized to one part of the structure but instead causes most or all of the structure to collapse, out of proportion to the original damage. This case led to a shift in philosophy in structural design. Before this attack, it was generally thought that special detailing of reinforced concrete construction was necessary only in areas of significant seismic hazard.

The Alfred P. Murrah Building was designed in the early 1970s based on ACI 318-71 concrete standard. It was constructed between 1974 and 1976. The main office structure was a nine-story reinforced concrete frame and shear wall building. It had a rectangular floor plan approximately 61 m (200 ft) long by 21.4 m (70 ft) wide. Three faces of the building were surrounded by other structures; one-story office buildings encompassed west and east sides and a multilevel parking structure was located on the south.

The structural system consisted of columns on a 6.1 x 10.7 m (20 x 35 ft) grid supporting deep floor beams spanning 10.7 m (35 ft) and a 150 mm (6 in.) one-way slab spanning 4.9 m (16 ft). To accommodate the architectural design the regular framing system was interrupted on the north elevation at the third floor level where every other column was terminated and supported by a wide 1.5 m (5 ft) deep transfer girder that spanned 12.2 m (40 ft) between two story columns. The second floor was also recessed and supported by a second deep beam spanning between short wall segments. The lateral force resisting system was principally composed of shear walls enclosing the elevator and stair core at the south side of the building. The ordinary reinforced concrete frame provided some additional lateral load resistance.

The truck bomb, estimated to be comparable to 1,800 kg (4,000 lbs) of TNT, was centered approximately 4 m (13 ft) from Column G20 of which only a small portion remained after the blast. Due to the lack of an alternative load path, upon the loss of support provided by column G20 four bays collapsed over the full height of the building. Additionally, upward pressure deflected the floor slabs and rotated the transfer girder at the third floor between column lines G16 and G26. Due to the good continuity between the floor slab and floor beams and the poor continuity between the floor beams and columns, loss of the floor slabs resulted in loss of the beams interconnecting and laterally bracing the columns and the transfer girder. One column buckled due to an overload and all floors and roof panels bounded by column lines 12, 28, F and G collapsed. In total 10 of 20 bays in the building footprint collapsed over the full height of the building. Approximately 4% or 543 m^2 (5850 ft^2) was destroyed directly by the blast while 42% or approximately 5,400 m^2 (58,100 ft^2) of the 12,700 m^2 (137,100 ft^2) was destroyed by blast and progressive collapse.

Lessons Learned

This collapse highlighted the need to develop details and procedures that would minimize the probability of progressive collapse. The best building performance is achieved with a complete three-dimensional structural system that interconnects all load path elements and provides stability and redundant load paths. Additionally, structural systems should incorporate mechanical fuses that allow elements to fail without destroying the entire system. Systems should be constructed to be strong and ductile to absorb overloads with large deformations and maintain continuity. Lower portions of perimeter columns should be designed to resist direct blast effects. Further, thorough investigation and analysis of collapses is essential to minimize unknowns and uncertainty on failure theories.

References

Corley, W.G., Mlakar, P.F., Sozen, M. A., and Thornton, C.H. (1998). "The Oklahoma City Bombing: Summary and Recommendations for Multihazard Mitigation," *Journal of Performance of Constructed Facilities*, 12(3), 100-112.

Delatte, Norbert J. (2009). *Beyond Failure: Forensic Case Studies for Civil Engineers*, ASCE Press, 155-162.

FEMA (1996). *The Oklahoma City Bombing. Improving building performance through multihazard mitigation*, FEMA 277, Washington, D.C.

FEMA (2005). *Blast-Resistance Benefits of Seismic Design- A Case Study*, FEMA 439A, Washington, D.C.

Hayes, J.R., Jr. et.al. (2005). "Can Strengthening for Earthquake Improve Blast and Progressive Collapse Resistance?" *J. of Structural Engineering*, 131(8), 1157-1177.

Hinman, E.E., and Hammond, D.J. (1997). *Lessons from the Oklahoma City Bombing, Defensive Design Techniques*, ASCE Press, New York, NY.

Mlakar, P.F., Corley, W.G., Sozen, M.A., and Thornton, C.H. (1998). "The Oklahoma City bombing: Analysis of blast damage to the Murrah Building," *Journal of Performance of Constructed Facilities*, 12(3), 113-119.

Osteraas, J.D. (2006). "Murrah Building Bombing Revisited: A Qualitative Assessment of Blast Damage and Collapse Patterns," *Journal of Performance of Constructed Facilities*, 20(4), 330-335.

Partin, B.K. (1995). *Bomb damage analysis of Alfred P. Murrah Federal Building, Oklahoma City, Oklahoma.* Report submitted to U.S. Congress, Alexandria, VA

Sozen, J.A., Thornton, C.H., Mlakar, P.F., and Corley, W.G. (1998). "The Oklahoma City Bombing: Structure and mechanisms of the Murrah Building," *Journal of Performance of Constructed Facilities.* 12(3), 120-136.

Wearne, P. (2000). *Collapse: When Buildings Fall Down*, TV Books, L.L.C. (www.tvbooks.com), New York.

CHARLES DE GAULLE AIRPORT TERMINAL 2E
(2004)

In the early morning hours of May 23, 2004, passengers in Terminal 2E at the Charles de Gaulle Airport in Paris reported loud noises that caused them to look up at the overhead concrete structure. What they saw, large cracks in the precast concrete shell structure, caused alarm leading to the evacuation of hundreds of passengers. Shortly after the initial discovery, a large section of the concourse collapsed resulting in several fatalities. The magnitude of the collapse at such a high profile public use building that had been open only 11 months led to a major investigation of Terminal 2E including the design process, construction sequences, and material properties.

Construction consisted of a combination steel truss on the exterior with a 30 cm (12 in.) thick precast concrete shell vault on the interior. A flattened area existed at the top of the vaults. The system was discontinuous at several locations due to openings for windows and skylights and intersecting corridors. The entire roof system was supported on a series of concrete column posts connected with concrete beams that were supported on sliding bearing connections. This section of Terminal 2E was 34 m (111 ft) wide and 650 m (2130 ft) long. Covering most of the exterior was a metal skin supported by steel framing and posts that was not rigidly attached in order to permit some thermal movement. Steel posts supporting the metal roof skin were recessed 10 cm (4 in.) into the precast shell sections.

As with many major failures in the historical record, technical experts investigating the collapse indicated a series of problems existed, both procedural and technical / structural as the cause of failure. Chief independent investigator, Jean Berthier, questioned the existence of proper design and construction processes in a project where the owner, project manager and architect were essentially the same. The terminal was designed by renowned French architect Paul Andreu, who was Director of Aeroports de Paris (AdP) at the time of the design. State owned AdP not only designed the building but they managed the construction, virtually eliminating the distinction between architect and client in this particular case.

Detailed modeling and study by the investigators concluded two possible trigger points of the collapse from the list of problems and deficiencies noted above. Berthier's team concluded that on the south side, the edge beam supporting the shell fractured, falling off its supports and collapsing. The other likely trigger mechanism was the punch through of several embedded steel struts which introduced high stresses and fatigue into the concrete shell partially as a result of thermal movement in addition to the existence of concrete creep stresses in the shells.

Reconstruction efforts for Terminal 2E consisted of a combination of demolition, modification and reconstruction of the roof and some related portions of the building.

Lessons Learned

In February of 2005, the results of a technical investigation commissioned by the French Minister of Transport indicated that the design process was not rigorous enough for a structure of this complexity and identified problems at various stages in the design process including:

- Lack of redundancy
- Inadequate or badly positioned reinforcing
- Steel support struts embedded too far into the concrete shell
- Weakened concrete shell support beams due to the passage of ventilation ducts
- Poor design and response to temperature variations that existed in the outer metal part of the structure

Beyond the technical items noted above, the lesson learned from this collapse is not new. The need for independent peer review, robust analysis, and more redundancy in unique structures that introduce new technology or combine technology in different ways has been recognized by the industry for years.

References

Brown, Jeff L. (2004). "Weakened Concrete May have Caused Paris Airport Collapse," *Civil Engineering*, September 14.

Conseil National des Ingenieurs et Scientifiques de France (2005). *"Synthese des travaux de la commission administrative sur les causes de l'effondrement d'une partie du terminal 2E de Paris-Charles de Gaulle,"* Paris, FR, Ministry of Transportation, Urban Design, Tourism, and Sea, February 15.

Downey, Claire (2005). "Investigation into Charles de Gaulle Terminal Collapse is Highly Critical," *Architectural Record News,* February 22. http://archrecord.construction.com/news/daily/archives/050222terminal.asp (accessed April 2, 2007).

Horn, Christian (2005). "Paris Air Terminal Collapse Report," *Architecture Week*, April 27. http://www.architectureweek.com/2005/0427/news_1-1.html (accessed April 2, 2007). http://www.equipement.gouv.fr/_article_print.php3?id_article=710 (April 2, 2007), (in French).

Leloup, Michele (2005). "Terminal de Roissy: Paul Andreu s'explique," *La Semaine France*, March 7, 2005.

Reina, Peter (2005). "Airport Roof Failure Blamed on Process," *Engineering News Record*, February 21, 10-11.

FOUR TIMES SQUARE SCAFFOLD COLLAPSE
(CONDE NAST TOWER)
(1998)

With virtually no warning, the exterior construction scaffold and personnel hoist on the exterior of the 48 story Conde Nast Tower in Times Square, New York partially collapsed. Starting at approximately 8:15 AM on July 21, 1998, a two story section of the scaffold at the 20th to 22nd floors buckled leaving almost 30 stories of scaffold above the collapsed area unsupported in the vertical direction and tilting precariously outward above the busy Times Square area of Manhattan while partial sections fell onto the streets and buildings below (Figure 5-21 and 5-22). A number of pieces of the steel and aluminum system fell onto buildings and the adjacent streets. An 18 meter (60-foot) length of the elevator hoist mast fell across 43rd Street, piercing the roof of an adjacent building and killing one occupant. Weighing over 5.44 tonnes (6 tons), the elevator counterweight fell to the street creating a crater 3 meters (10 ft) deep in the pavement. Although both cabs of the personnel hoist were in operation and occupied by construction workers, both groups were able to discharge at different levels just before the collapse. One cab operator, who went back for his tools, had the collapse occur around him and was saved only because he was able to grab a piece of rope and was pulled to safety by others.

The Conde Nast Tower stretches the full block from 42nd Street to 43rd Street. The initial failure and potential for additional collapse presented an ongoing danger to pedestrians, vehicle occupants, and those individuals in the buildings in the Times Square area. New York City Office of Emergency Management was summoned to the site within minutes of the collapse and a crisis management team was quickly formed consisting of a number of city agencies as well as private consultants. LZA Technology, a division of the Thornton-Tomasetti Group was retained to evaluate the collapse conditions, contain the disaster, create an emergency stabilization plan, and design and implement the demolition process for the failed structure. Wiss Janney Elstner (WJE) was retained by the New York City Department of Buildings to investigate and determine the cause of the failure.

Lessons Learned

Results of the investigation indicated that the mechanism of the failure was buckling of the scaffold frame legs just below the 21st floor. Structural analysis led investigators to conclude that it was missing bracing (design condition compared to the as-built condition) that caused the aluminum scaffold frame legs to fail in buckling. Based on structural analysis, the missing bracing was determined to have reduced the vertical load capacity of the scaffold in the area of collapse by approximately 34%. Investigators concluded that the required bracing, necessary for stability, had never been installed.

References

Board of Inquiry of the City of New York. (1999). *Four Times Square, Board of Inquiry Report*, September 29.

Dunlap, David W. (1998). "Construction Collapse in Times Square: The Repair; Giant Net to Enclose Scaffold To Keep Debris From Streets," *The New York Times*. Available at http://query.nytimes.com/gst/fullpage.html?res=9C02E2DF1539F930A15754 C0A96E958260&n=Top%2fReference%2fTimes%20Topics%2fOrganization s%2fD%2fDurst%20Organization, Archives, nytimes.com, (accessed May 11, 2007)

Engineering New Record (ENR). (1998). "Causes Probed as Crippled Hoist Comes Slowly Down," *ENR*, McGraw-Hill, August 10.

ENR. (1998). "No Single Trigger in Hoist Failure," *ENR*, McGraw-Hill, August 10.

Rentschler, Glenn P., and Walkup, Stephanie, (2000). "The Times Square Scaffold Collapse," *Forensic Engineering: Proceedings of the Second Congress*, Technical Council on Forensic Engineering, ASCE.

The Council on Tall Buildings and Urban Habitat – *Emporis*. Available at http://www.emporis.com/en/wm/bu/?id=condenastbuilding-newyorkcity-ny-usa, (accessed on May 11, 2007).

Velivasakis, Emmanuel E. (2000). "4 Times Square Hoist Collapse, The Handling of a Potential Disaster," *Proceedings of the Second Congress, Technical Council on Forensic Engineering*, ASCE.

Wiss, Janney, Elstner. (1999). *Investigation of the Partial Collapse of the Personnel Hoist Scaffold System at the Conde Nast Building*, WJE, Princeton, NJ, July 1.

Figure 5-21. Collapse of scaffolding of Four Times Square.
Source: Courtesy of Daniel A. Cuoco (Thornton Tomasetti).

Figure 5-22. Collapse of scaffolding of Four Times Square.
Source: Courtesy of Daniel A. Cuoco (Thornton Tomasetti).

Index

Page numbers followed by *f* indicate figures.

ACI Building Code, 77
Alcan Highway, 61
Alfred P. Murrah Federal Building, 116–117
AMC Warehouse, 77–78, 78f
anchorages, 61
Andreu, Paul, 118
Antelope Valley Freeway Interchange, 66–67, 68f
Ashtabula Bridge, 50–51

bearing capacity, 5, 7
Binghamton State Office Building, 98–99
bridge failures: Antelope Valley Freeway Interchange, 66–67, 68f; Ashtabula Bridge, 50–51; Cypress Viaduct, 74–75, 75f; Falls View Bridge, 56; King Street Bridge, 63; Mianus River Bridge, 69–70; Peace River Bridge, 61; Point Pleasant Bridge—Silver Bridge, 64–65; Quebec Bridge, 54–55; Sando Arch, 57; San Francisco-Oakland Bay Bridge, 71–72, 72f, 73f; Schoharie Creek Bridge, 9–10, 11f–13f; Second Narrows Bridge, 62; Tacoma Narrows Bridge, 58–60; Tay Bridge, 52–53
brittle fracture failure, 63
building failures: Alfred P. Murrah Federal Building, 116–117; AMC Warehouse, 77–78, 78f; Binghamton State Office Building, 98–99; Burnaby Supermarket Rooftop Parking Deck, 110–111; California State University, Northridge, Oviatt Library, 114–115; Charles de Gaulle Airport Terminal 2E, 118–119; Four Times Square Scaffold Collapse (Conde Nast Tower), 120–121, 122f; Hartford Civic Center Coliseum, 88–90, 90f–92f, 95; Hyatt Regency Hotel Pedestrian Walkways, 95, 100–101, 102f, 103f; Imperial County Services Building, 93, 94f; Kemper Memorial Arena Roof, 95–96, 97f; L'Ambiance Plaza, 106–107, 108f, 109f; Northridge Meadows Apartments, 112, 113f; Pino Suarez Building, 104–105, 105f; Ronan Point Tower, 79–80, 81f, 82f; Skyline Plaza Apartment Building (Bailey's Crossroads), 85–86, 87f; 2000 Commonwealth Avenue, 82–83, 84f
Bureau of Public Roads, 61
Burnaby Supermarket Rooftop Parking Deck, 110–111

California Regional Water Quality Control Board, 38
California State University, Northridge, Oviatt Library, 114–115
Carsington Embankment, 32
Center for Building Technology (National Bureau of Standards), 85
Charles de Gaulle Airport Terminal 2E, 118–119
Comprehensive Environmental Response, Compensation and Liability Act (1980), 34, 36, 38, 40
Conde Nast Tower, 120–121, 122f
control dredge construction methods, 30
Cryptosporidium parvum, 45, 46
Cypress Viaduct, 74–75, 75f

dam failures: Lower San Fernando Dam, 22–23, 23f; Malpasset Dam, 17–18; St. Francis Dam, 15–16; Teton Dam, 24–25, 25f–28f; Vajont Dam, 19–20, 21f
deflector vanes, 59

earthquakes: bridge damage, 71–72, 72f, 73f; building collapse, 93, 94f, 104–105, 105f, 112, 113f, 114–115; dam slide, 22; freeway collapse, 66–67, 68f; predictive capabilities, 8

electrical transformer fires, 98–99
embankment construction, 32
environmental failures. *See* geoenvironmental failures
Environmental Protection Agency (EPA), 38, 98

Falls View Bridge, 56
Fargo Grain Elevator, 7
Farquharson, F. B., 58
Federal Works Agency, 59
fire, chemical contamination due to, 98–100
foundation failures: Fargo Grain Elevator, 7; La Playa Guatemala Earthquake, 8; Schoharie Creek Bridge, 9–10, 11f–13f; Tower of Pisa, 2–3, 4f; Transcona Grain Elevator, 5–6
Four Times Square Scaffold Collapse (Conte Nast Tower), 120–121, 122f

geoenvironmental failures: Kettleman Hills Waste Landfill, 42–43, 44f; Love Canal, 34–35; North Battleford, Saskatchewan Water Treatment Failure, 45–47, 48f; Seymour Recycling Facility, 40–41; Stringfellow Acid Pits, 38–39; Valley of the Drums, 36–37
grain elevators: Fargo, 7; Transcona, 5–6

Hartford Civic Center Coliseum, 88–90, 90f–92f, 95
hazardous wastes: Kettleman Hills Waste Landfill, 42–43, 44f; Love Canal, 34–35; Seymour Recycling Facility, 40–41; Stringfellow Acid Pits, 38–39; Valley of the Drums, 36–37

Hooker Chemical and Plastic Company, 34
Hyatt Regency Hotel Pedestrian Walkways, 95, 100–101, 102f, 103f

I-bar suspension bridge, 64
ice jam, 56
Imperial County Services Building, 93, 94f
Imperial Valley Earthquake (1979), 93
Interstate 5/14 Freeway collapse, 66, 67, 68f

Kemper Memorial Arena Roof, 95–96, 97f
Kentucky Department of Natural Resources and Environmental Protection (KDNREP), 36
Kettleman Hills Waste Landfill, 42–43, 44f
King Street Bridge, 63

L'Ambiance Plaza, 106–107, 108f, 109f
landfill, Kettleman Hills Waste Landfill, 42–43, 44f
landslide, Rissa Norway Landslide, 29
La Playa Guatemala Earthquake, 8
Leu, Sherwood, 62
liquefaction, earthquake-induced, 8
Loma Prieta Earthquake (1989), 71
Love, William T., 34
Love Canal, 34–35
Lower San Fernando Dam, 22–23, 23f
LZA Technology, 120

MacDonald, Charles, 50
Malpasset Dam, 17–18
Mexico City Earthquake (1985), 104
Mianus River Bridge, 69–70
Moisseiff, Leon, 58
Mulholland, William, 15

National Bridge Inspection Standards (NBIS), 64
National Bureau of Standards (MBS), 85, 100

National Priorities List, Seymour
Recycling Corporation, 40
Nerlerk Berm Failure, 30–31
New Mexico State Highway
Department, 98
New York City Office of Emergency
Management, 120
Niagara Falls, New York, 34, 56
Niagara Power and Development
Company, 34
North Battleford, Saskatchewan Water
Treatment Failure, 45–47, 48f
Northridge Earthquake (1994), 66–67,
112, 114–115
Northridge Meadows Apartments, 112,
113f

Occupational Safety and Health
Administration (OSHA), 85

PCBs (polychlorinated biphenyls), 98
Peace River Bridge, 61
Pino Suarez Building, 104–105, 105f
Point Pleasant Bridge—Silver Bridge,
64–65
prefabricated construction techniques,
79
Public Works Administration, 58

Quebec Bridge, 54–55
quick clay construction, 29

Rainbow Arch, 56
Reconstruction Finance Corporation,
58
Rissa Norway Landslide, 29
Ronan Point Tower, 79–80, 81f, 82f
roof collapse, 95–96, 97f
rooftop parking deck collapse, 110–
111

sand berm construction, 30
Sando Arch, 57
San Fernando Dam, 22–23, 23f
San Fernando Earthquake (1971), 22

San Francisco-Oakland Bay Bridge,
71–72, 72f, 73f
Saskatchewan Environment and
Resource Management (SERM), 45
scaffold collapse, 120–121, 122f
Schoharie Creek Bridge, 9–10, 11f–
13f
scour, effects of bridge, 9
Second Narrows Bridge, 62
Severn-Trent Water Authority, 32
Seymour Recycling Corporation, 40
Seymour Recycling Facility, 40–41
Skyline Plaza Apartment Building
(Bailey's Crossroads), 85–86, 87f
slope movement, Vajont Dam, 19–20
soil bearing capacity, 5
soil-extraction method, 2–3
soil mechanics, 5
St. Francis Dam, 15–16
State University of New York at New
Paltz, 98
steel selection, 63
Stone, Amasa, 50
Stringfellow Acid Pits, 38–39
Stringfellow Quarry Company, 38
suspension bridge failure, 64

Tacoma Narrows Bridge, 58–60
Tay Bridge, 52–53
Taylor, A. L., 36
Teton Dam, 24–25, 25f–28f
Thornton-Tomasetti Group, 120
Tower of Pisa, 2–3, 4f
Transcona Grain Elevator, 5–6
2000 Commonwealth Avenue, 82–83,
84f

underwater sand berm construction, 30
University of Washington at Seattle, 59
U.S. Army Cold Regions Research and
Engineering Laboratory, 88

Vajont Dam, 19–20, 21f
Valley of the Drums, 36–37

walkway collapse, 100–101, 102f,
    103f
Washington Toll Bridge Authority, 58
water treatment facility, North
    Battleford, Saskatchewan Water

Treatment Failure, 45–47, 48f
Wilkins Air Force Depot, 77
wind loading, 95
wind tunnel testing, 59
Wiss Janney Elstner (WJE), 120